W0079261

Hospitality, Home and Life in the Platform Economies of Tourism

Maartje Roelofsen

Hospitality, Home and Life in the Platform Economies of Tourism

palgrave
macmillan

Maartje Roelofsen
Faculty of Economics and Business
Universitat Oberta de Catalunya
Barcelona, Spain

ISBN 978-3-031-04009-2   ISBN 978-3-031-04010-8   (eBook)
https://doi.org/10.1007/978-3-031-04010-8

© The Author(s), under exclusive license to Springer Nature Switzerland AG 2022
This work is subject to copyright. All rights are reserved by the Publisher, whether the
whole or part of the material is concerned, specifically the rights of translation, reprinting,
reuse of illustrations, recitation, broadcasting, reproduction on microfilms or in any other
physical way, and transmission or information storage and retrieval, electronic adaptation,
computer software, or by similar or dissimilar methodology now known or hereafter
developed.
The use of general descriptive names, registered names, trademarks, service marks, etc.
in this publication does not imply, even in the absence of a specific statement, that such
names are exempt from the relevant protective laws and regulations and therefore free for
general use.
The publisher, the authors and the editors are safe to assume that the advice and information in this book are believed to be true and accurate at the date of publication. Neither
the publisher nor the authors or the editors give a warranty, expressed or implied, with
respect to the material contained herein or for any errors or omissions that may have been
made. The publisher remains neutral with regard to jurisdictional claims in published maps
and institutional affiliations.

Cover illustration: © John Rawsterne/patternhead.com

This Palgrave Macmillan imprint is published by the registered company Springer Nature
Switzerland AG
The registered company address is: Gewerbestrasse 11, 6330 Cham, Switzerland

Acknowledgements

This book relies crucially on the contributions of numerous *Airbnb* hosts in Bulgaria, Denmark, Ghana, Italy, Spain, and the Netherlands. I first wish to express my heartfelt appreciation for their dedication to my research project over the years. Many of these hosts have shared their homes and everyday life with me for periods of days and weeks, and have shared with me their expertise and experiences. In addition, I have had the privilege to rely on the amazing support of numerous people who selflessly enabled my reading, writing, and thinking in different ways:

Donna Houston, Julie Wilson, Karin Peters, Lluís Garay Tamajón, René van der Duim, Soledad Morales Pérez, and Ulrich Ermann. Your support and belief in my capacities as a scholar have meant a great deal to me and I am incredibly appreciative for all that you have done for me in the past years. I feel indebted to each one of you. Heide Bruckner and Madeline Donald, I have been so lucky to find in both of you a dear friend, an emotional support, and an intellectual inspiration during my years as a doctoral- and early career researcher. You have kept my moral up and always encouraged me to consider even more interesting directions to take! Kiley Goyette, I am in awe of your work and can only hope that I will be able to read more of it. Thank you so much for joining me in questioning the divisions of *Airbnb* work—one of the most rewarding research endeavours I have experienced to date!

Thank you, Tanja Zöhrer and Farhana Haque for assisting me in all the formalities and for making Austrian and Australian bureaucratic procedures a little less tense; Michael Kolesnik and Daniel Blazej, for having so richly illustrated this book with maps; Simon Würzler, Marlies Bodinger, and Radostina Zasheva, for your tremendous support during the collection of data in Bulgaria and Graz. Thank you, Ben, Ben, Ben (Reza), Charlotte, Eefje, Emmelyn, Reinier, and Ria for contributing to my overall wellbeing and for giving me that final push towards submission. Alessandro, Amos, Bob, Christina, Coen, Elsa, Jelmer, Laura, Marco, Mirjam, Mohamed, Rita, Sander, Tanya, and Vera. Your friendship has given me comfort, energy, and always placed me back firmly with my two feet on the ground when needed. Very special thanks go out to Claudio. You have managed to ease me out of my existential and intellectual comfort zone very elegantly and have been an amazing writing partner over the years. The completion of this book owes a lot to your ongoing support, and several of the chapters that feature in this book are based on our fruitful collaborations. We have come to complement each other well and I look forward to our future research projects. Thank you so much for your patience, and for taking care of me (and the home!) while I was working evenings and weekends in the past years to complete this project.

Finally, I wish to thank the Universitat Oberta de Catalunya for funding my postdoctoral stay and for providing me with all the support to finish writing this book. My visit to Bulgaria and related fieldwork at the time was supported by the University of Graz Doctoral Stipend and the 2015 Rudi Roth Grant, for which I wish to express my gratitude. Many thanks to the Geography Division and the Department of History and Cultures at the University of Bologna for taking me on as a Visiting Scholar throughout the early months of the pandemic. Your institutional support has been fundamental to keeping this project going.

Various arguments, sections, and chapters throughout this book have been drawn from previous published journal articles and chapters. Specifically:

Thanks to the editors of *Oikonomics* and the Open University of Catalunya for permission to use parts of "Sanitized homes and healthy bodies. Reflections on Airbnb's response to the pandemic" originally published in *Oikonomics*, 2021, 15 (May). https://doi.org/10.7238/o.n15.2104 (Chapters 2 and 3)

Thanks to the editors of *Erdkunde* and the University of Bonn for permission to reproduce parts of "Exploring the Socio-Spatial Inequalities of Airbnb in Sofia, Bulgaria" originally published in *Erdkunde*, 2018, 72 (4): 313–27. https://doi.org/10.3112/erd kunde.2018.04.04 (Chapter 3)

Thanks to the editors of *Fennia* and the Geographical Society of Finland for permission to reproduce parts of "Performing 'Home' in the Sharing Economies of Tourism: The Airbnb Experience in Sofia, Bulgaria" originally published in *Fennia—International Journal of Geography*, 2018, 196 (1): 24–42. https://doi.org/10.11143/fen nia.66259 (Chapter 4)

Thanks to Claudio Minca, the editors of *Geoforum* and Elsevier for permission to reproduce parts of "The Superhost. Biopolitics, Home and Community in the Airbnb Dream-World of Global Hospitality" originally published in *Geoforum*, 2018, 91: 170–81. https://doi. org/10.1016/j.geoforum.2018.02.021 (Chapter 5)

Thanks to Claudio Minca, the editors of *Tourism Geographies* and Taylor and Francis for permission to reproduce parts of "Becoming Airbnbeings: On Datafication and the Quantified Self in Tourism" originally published in *Tourism Geographies*, 2019, 23 (4): 743–64. https://doi.org/10.1080/14616688.2019.1686767 (Chapter 5)

CONTENTS

LIST OF FIGURES

CHAPTER 1

Introduction

Abstract This introduction chapter provides an outline of the structure of the book. Drawing on the story of Freya, a short-term rental host in Barcelona, Spain, it introduces the idea of the "platform economy" within the context of tourism. It concludes with some notes on the various methods that have been used in the research that underpins the book.

Keywords Platforms · Tourism · Hospitality · Platform economy · *Airbnb*

Freya doses off while watching her favourite TV show, as she often does towards the end of the afternoon. She lies comfortably on the sofa in the living room of her 3-bedroom apartment in the city centre of Barcelona, Spain. As I enter through the front door, I make my presence noticed by uttering a high-pitched "Hola" to override the TV volume. Freya replies with a "Hola, que tal?" and I pass through the hallway to enter the bedroom, shaking off my bags and the fatigue of the day. It has become our pattern to have a small chat about our day while I heat up my dinner in the kitchen. Today was a difficult day for Freya since the hired help called off sick, pressing her to unexpectedly do the weekly cleaning of the apartment herself. The kitchen and bathroom in the apartment are the only two common spaces that Freya shares with me and

© The Author(s), under exclusive license to Springer Nature Switzerland AG 2022
M. Roelofsen, *Hospitality, Home and Life in the Platform Economies of Tourism*, https://doi.org/10.1007/978-3-031-04010-8_1

Fiona, another guest who has been staying with Freya for a few months. Freya wishes to keep the living room to herself, and she has provisionally and partially fenced it off with a folding screen that has "private" written on it, suggesting it is off limits to her guests. This explicit bordering practice is commonplace in many households that I stayed with over the years while carrying out the research that underpins this book. Some see the demarcation of "private space" through objects and practices as an entitlement that comes with their job as "keepers of their inn". Freya is, in fact, a "host" who rents out two bedrooms through two different short-term rental *platforms*.

Platforms have been defined as digital architectures and infrastructures that facilitate and organize economic activity, social exchange, political debate and wide range of other societal activities and functions (see Leszczynski 2020). Often reliant on networking technologies, platforms tend to bring together different user groups—like hosts and guests— and allow for the production and exchange of dynamic, user-generated content. Content that users freely and sometimes unknowingly provide to the platform in the form of visual, textual, numerical, and geolocational data. Freya's profile on the platform, for example, features extensive written and photographic descriptions of her apartment and the bedrooms that she rents out, as well as a description of herself. It features a portrait photo of Freya, her biography, numerous ratings and reviews on her home and hospitality submitted by guests who stayed with her, and a Superhost status badge, which she has been assigned for having provided a distinguished level of hospitality over time.

Within the context of tourism, platforms have functioned as "brokers" in digitized marketplaces where accommodation, hospitality, transport, and a huge array of other products and services are exchanged among people. Their online architectures facilitate and control a wide range of activities, some collaborative and communal, others competitive and profit driven. These economies encompass a plethora of organizational forms, many of which mirror "traditional" capitalist relations of production; commodifying everyday spaces, resources, and labour that were previously considered "unproductive". What commonly unites these socio-technological formations is their role in mediating forms of exchange and deriving value from that what is being circulated between the users of these platforms (Ince and Hall 2018).

Many of the guests who stay with hosts like Freya are tourists looking to explore a city or town for a few days, while having the comforts of residential housing that hotels, motels, or hostels lack. Others, like Fiona, are students or mobile (knowledge) workers from afar, committed to staying for a few days, weeks, or months while carrying out their duties. Over the years, Freya has connected through the platforms' software with whomever she thinks has a suitable profile—usually white, middle-class women who can afford to pay the rent upfront. She operates her business on certain biases, and she is selective because she can. The platform allows her to decline requests from prospective guests whose profiles, expectations, and demands do not suit her. To her advantage, there is an abundance of demand for rooms in her neighborhood, so she is never shy of new inquiries whenever she declines. And like many other hosts whom I interviewed over the years, Freya already receives a monthly income— in this case, a pension—thus the money that she earns by renting out the rooms is not crucial to her survival but allows her to maintain her comfortable middle-class lifestyle. A lifestyle and socio-economic status she brought with her when she migrated to Barcelona from abroad, many years back. Being a fervent traveller herself, she speaks three languages fluently and she enjoys chatting with the "comfortably exotic" women who stay in her home (Ladegaard 2018) and who provide her with companionship as a divorcee now living alone.

Freya's tech-savvy adult children who live elsewhere have helped her set up an online profile on these platforms some years ago and have re-decorated the rooms to make them fit for rent to international guests. The rooms radiate a personalised "Scandinavian" hipster aesthetics, combining IKEA furniture with some of Freya's personal items (Myambo 2021); a certain middle-class décor propagated in the home styling guides that the short-term rental platforms provide to their hosts (Bialski 2017). Freya has been a host on these platforms for several years now. It has changed her relationship with her home, which has at once become a sphere of production *and* social reproduction. She earns an income from the spaces and the (outsourced) household labour that previously only had value in the sustenance of her own daily life and that of her family. By turning her home into a workspace, housework and emotional labour have come to have different values, especially now that these aspects are reviewed, rated, and paid for by her guests. The personal relationships that Freya develops with her guests are mostly ephemeral but that does not prevent her from working hard to provoke feelings of homeliness and comfort among them. Yet, paradoxically, Freya's guests also sense that they have transgressed a space that has a deep and intimate meaning to Freya. Like Fiona, I make sure not to occupy the common spaces for longer than needed to prepare and eat my meals and we always clean up after ourselves. At the same time, Freya admits that while her newfound economy gives her a sense of accomplishment as a pensioner, having short-term guests at home could never quite replace the kind of familial life she used to have a time long gone. Moreover, with the onset of the COVID-19 pandemic she has struggled to outweigh the benefits and drawbacks of hosting, as she constantly faces new encounters with potentially infectious guests. Even more so than before, the health crisis has forced her to reassess more carefully the meaning of life and home in its contribution to ontological security and wellbeing.

Freya is certainly not the only short-term rental host in Barcelona, and not even the only one in her 5-storey apartment building. In fact, short-term rentals have proliferated in Barcelona to such an extent that they have formed a serious threat to the availability and affordability of long-term rental housing in the city. Freya's motivations to host and her residential circumstances are common among other platform users but not necessarily reflective of the status quo renting out on these platforms. Over the years, short-term renting has become an incredibly lucrative business opportunity for groups of investors and speculators who purposely buy properties to be rented out in their entirety on these platforms, as they receive much higher rents from tourists than they would from regular tenants. Freya is aware of the public resistance against short-term rentals and keeps up with the political debates and attempts at regulating this activity in Barcelona (Wilson et al. 2021). She also recounts her experience of the social unrest that started to emerge a decade ago when Barcelona became a poster child for all the harmful excesses that mass tourism had provoked. Despite this knowledge, Freya has remained committed to her newfound economy and resulting lifestyle and has no serious intention to stop.

This book builds upon the premise that today's socio-spatial relationships cannot be fully understood without acknowledging the role of the digital (Lupton 2016; Ash et al. 2018). It argues that the digital *enables* but also actively *constitutes* specific practices of hospitality and *produces* specific intimate spatialities in people's homes. As intermediaries, digital short-term rental platforms not only facilitate exchanges but also shape the social dynamics that depend on them (Gillespie 2015, p. 2). Their technical, economic, and political design significantly influence what people are (and are not) able do in that specific domain determined by the platform (ibid.). Through algorithms platforms operationalize their own rules, which consequently promote, select and qualify some data and content over other. My stay with Freya, for example, was not accidental: it was the result of a specific search query that prompted algorithms to sort through geolocalized data provided by thousands of hosts just like Freya. It provided me with hundreds of results on a map-based interface and an accompanying ranking of listings in Barcelona, the order of which likely should have appealed to my preferences for a specific neighbourhood, type of room and the amount of overnight fees I could afford. While users of these platforms, like myself, may be aware that these algorithms exist and shape their interactions, most users know very little about *how* these regulators of connections and exchange actually work (Bucher 2012; Gillespie 2015). This makes studying platforms all the more important and it has provoked me to further reflect on the ways in which data is collected and used to shape people's routines, behaviours, and practices. It has led me to question how algorithmic management increasingly gets entangled with everyday spaces and affects how we travel, where we drive, how we act as "local residents", and to which unknown guests we open our doors? Moreover, how do users of these digital infrastructures hold platform enterprises to account and resist algorithmic decision-making? I hope to answer some of these questions throughout the book.

Freya's story is one out many, but it illustrates how the city, the home, and everyday life are inextricably linked through short-term rental platforms; they affect people's lives at every geographical scale, whether they are direct participants in these platform economies or not. This is the main thread that runs through this book, which is outlined as follows. In the second chapter, I begin by tracing some of the earliest digital developments in tourism that have provided the foundations of, what are perceived today, the most "revolutionary" platform economies in tourism. I do so to explicate that these platforms do not exist in a vacuum but have extended from other technological innovations which, for the most, were similarly intended to automate, modernize, and scale the distribution of information and sales, eventually contributing to the unfettered growth of global mobility, tourism, and hospitality. After reflecting on some of the working definitions of the term "platform" the chapter continues to show how various tourism platforms at once challenge and replicate existing political-economic modes of organization. I do so by questioning the commonly made association of platforms with the "sharing economy", a "new" economic model that grew out of a certain political-economic climate generated by the global financial crisis of 2007–2008. After giving a more detailed account of the numerous and diverse platform economies involved in the business of tourism and hospitality, the final sections of this chapter serve to provide some background on *Airbnb*, the platform that acts as the main subject of research throughout most of the remainder of the book.

Chapter 3 provides insight into the various socio-spatial and economic impacts that short-term rental platforms have had on many cities, neighbourhood residents and the hotel sector. It argues that the conditions that allow platforms to flourish are contingent upon the different political and sectoral contexts in which they operate, the public interests that are at stake, as well as governments' interest and ability to enforce (new) regulations. By means of illustration, the chapter describes the development of *Airbnb* in Sofia, Bulgaria and sheds light on the broader urban transformations that have taken place in the city since the end of the socialist regime in 1989. It provides empirical evidence of the unevenly distributed economic benefits produced by *Airbnb*, further adding to existing socio-economic disparities within the context of Sofia.

In chapter 4, I zoom in on the level of the household to explore how short-term rental platforms transform the intimate spatialities, everyday practices, and the social relations that intersect at "home". Drawing again on *Airbnb* within the context of Sofia, the chapter shows that the meaning of home is continuously reshaped by the social and emotional relationships that are established between different hosts and different guests. While being disruptive to some households the chapter also argues that the *Airbnb* economy represents an opportunity for some hosts to produce and extract new values from their intimate spatialities and their ordinary practices of homemaking.

Chapter 5 explores the effects of *Airbnb*'s operations and politics at the scale of individual life and the body. It draws attention to the tensions that emerge between the platform's rhetoric and its fixation on the extraction of value from the everyday lives of hosts and guests. The chapter analyses several sites of evaluation at the core of the *Airbnb* machinery, unravelling the technologies and the calculative rationalities that underlie the "becoming of" the *Superhost*, a certain status that hosts are deemed to aspire, or, at least, are nudged to do achieve. The chapter also shows how spaces of hospitality are qualified through the systematic incorporation of intimate relations between the bodies of strangers, and how these relations shape new understandings of travel, community, and "home".

A Note on Methods

The different scales of analysis adopted in this book have required three different approaches associated with a different set of methods, which I explain in more detail in the respective chapters. The broader urban geographies of *Airbnb* in Sofia that I describe in Chapter 3 have been studied using a qualitative web content analysis followed by a detailed examination of all the *Airbnb* listings available in the city during different periods of time. The analysis of *Airbnb* "homes" in Chapter 4 is instead based on a set of ethnographic methods, including interviews, participant observation as well as autoethnography, which required long and intense periods of research in Sofia, where I stayed as an *Airbnb* guest. These methods have been employed in line with the broader analytical framework provided by performance theory. The performance of "home" and the related qualification and ranking of such performances are in fact a key element emphasized by the platform and its workings. In Chapter 5 biopolitical analytics has been adopted in order to scrutinize the ways in which the platform qualifies and capitalizes on the ordinary lives of its hosts. Here, the "digital lives" of *Airbnb* users have been studied via a qualitative analysis of the *Airbnb* website, affiliated blogs, promotional material, hospitality guidebooks, blog posts, discussion boards, and policies which included both user-generated content as well as material published by the platform itself. Moreover, the platform's exchange mechanisms, structuring logic, and key algorithms were analysed to understand how indeed lives in this "sharing" economy were qualified.

Finally, *Airbnb*'s role in the organization of social and spatial practices of hospitality in everyday life was explored by taking on the role of host myself. This has offered an opportunity to familiarize myself with the digital infrastructure of *Airbnb* from a host's perspective. In addition, it has represented a useful way to experience the embodied and emotional textures of hosting. I have conducted a pilot study in Bennekom, the Netherlands, where I made use of a holiday cottage that was owned, at the time, by one of my family members. This property was advertised through the *Airbnb* platform in accordance with local laws and regulations on holiday rentals. From March to December 2016, I hosted a total 23 stays during which the entire property was rented out. As a host, I attempted to fulfil all the requirements that *Airbnb* ascribes to "Superhosts" by committing myself to the many practices that make up the "ideal hosting" (e.g. communicating on- and offline, cleaning, welcoming in person, remaining available, taking care of the financial administration, etc.). During this period, I gained an in-depth understanding of the amount of time and (new) kinds of labour involved in providing an "*Airbnb* experience". By being an *Airbnb* host, a comprehensive set of hosting "tools" became available to me on the platform. For example, I was encouraged to "self-track" my performances as a host through the *Airbnb* "Dashboard", which incorporates metrics applications that monitor certain on- and offline practices and provided statistical averages on a range of achievements. By interacting with these tools, I have gained a deeper understanding of how the platform monitors its users, and how it nudges them to perform better in line with the platform's standards. Although the specificities of the advertised property and its location have shaped my experiences as a host, the aim of the pilot study was not to say anything specific about Bennekom or *Airbnb*'s operations in Bennekom. The Bennekom pilot study is only indirectly referred to in Chapter 5, where I examine the many sites and systems of data collection and the processes of data-driven forms of governance by the *Airbnb* platform.

REFERENCES

Ash, James, Rob Kitchin, and Agnieszka Leszczynski. 2018. Digital Turn, "Digital Geographies?" *Progress in Human Geography* 42 (1): 25–43. https://doi. org/10.1177/0309132516664800.

Bialski, Paula. 2017. "Home for Hire: How the Sharing Economy Commoditises Our Private Sphere." In *Sharing Economies in Times of Crisis*, edited by Anthony Ince and Sarah Marie Hall, 83–95. Routledge.

Bucher, Taina. 2012. "Want to Be on the Top? Algorithmic Power and the Threat of Invisibility on Facebook." *New Media and Society* 14 (7): 1164–1180. https://doi.org/10.1177/1461444812440159.

Gillespie, Tarleton. 2015. "Platforms Intervene." *Social Media and Society* 1 (1). https://doi.org/10.1177/2056305115580479

Ince, Anthony, and Sarah Marie Hall. 2018. *Sharing Economies in Times of Crisis: Practices Politics and Possibilities*. London: Routledge.

Ladegaard, Isak. 2018. "Hosting the Comfortably Exotic: Cosmopolitan Aspirations in the Sharing Economy." *The Sociological Review* 66 (2): 381–400. https://doi.org/10.1177/0038026118758538.

Leszczynski, Agnieszka. 2020. "Glitchy Vignettes of Platform Urbanism." *Environment and Planning D: Society and Space* 38 (2): 189–208. https://doi. org/10.1177/0263775819878721

Lupton, Deborah. 2016. *The Quantified Self*. Cambridge, UK: Polity.

Myambo, Melissa Tandiwe. 2021. "Glocal Hipsterification: Hipster-Led Gentri-fication in New York's, New Delhi's, and Johannesburg's Cultural Time Zones." In *Hipster Culture: Transnational and Intersectional Perspectives*, edited by Heike Steinhoff, 47–64. Bloomsbury Publishing USA.

Wilson, Julie, Lluís Garay-Tamajon, and Soledad Morales-Perez. 2021. "Politi-cising Platform-Mediated Tourism Rentals in the Digital Sphere: Airbnb in Madrid and Barcelona." *Journal of Sustainable Tourism*, 1–22. https://doi.org/10.1080/09669582.2020.1866585.

CHAPTER 2

A Brief History of Tourism Platforms

Abstract It would be tempting to read the onset of tourism platforms as sudden and emblematic developments of their time. Yet, the tourism industry has long been considered one of the early adopters of computerized systems and digital technologies to automate transactions, to accelerate certain work processes and to optimize scale production. This chapter provides a brief history of the intimate link between tourism and the digital. In doing so, it shows how today's tourism platforms as techno-organizational formations are both novel as well as conventional in their replication of already existing modes of political economic organization. In the latter sections of this chapter, the brief history and digital architecture of *Airbnb* is described since it features as a case study throughout the remainder of the book.

Keywords History · Digitalization · Tourism · Platforms · *Airbnb*

THE DAWN OF COMPUTER RESERVATION SYSTEMS

Historical accounts of the digitalization of tourism often start in the late 1950s and 1960s with the invention and implementation of the first Computer Reservation System (CRS) in the aviation sector (Buhalis and Law 2008). The invention of the CRS has been celebrated as one of the

© The Author(s), under exclusive license to Springer Nature Switzerland AG 2022
M. Roelofsen, *Hospitality, Home and Life in the Platform Economies of Tourism*, https://doi.org/10.1007/978-3-031-04010-8_2

11

key moments in the history of the tourism and travel industry that—jointly with the arrival of commercial jet engine planes in the 1950s—has led to the phenomenal expansion of tourism mobility (Yano 2011). Jet travel in those early years was successfully marketed and advertised to the Euro-American public as the ideal of modernity, a techno-utopic fantasy. By the end of the 1960s, air travel had become an activity no longer reserved for the elites, although it remained notably segregated along social class, urban or rural status, race and ethnicity (ibid.). Between the 1950s and 1970s, an average increase of global air traffic passengers of over 10% a year (Oxley and Jain 2015) made airline schedule planning increasingly complex, and manual reservation management processes had reached massive proportions. Updating airline inventories and processing passenger reservations were laborious and time-consuming procedures that involved manually administering paper record cards and communication through telephones and telexes (Truitt et al. 1991). Timetables of flights and pricing of airline seats were still advertised through printed newspapers at the time. To make the distribution of airline information and sales more efficiently, the objective was thus to create a computer system that would be able to execute all related processes automatically. Research that led to the development of one of the first CRSs was initiated through a collaboration between US-based American Airlines and IBM, the International Business Machines Corporation (ibid.). The computerized data processing system that emerged from this collaboration between American Airlines and IBM was coined the *Semi-Automated Business Research Environment* (SABRE), which became fully operational in 1964. It could make American Airlines' data on airline seats and passengers electronically available across the U.S. and Canada by connecting 1500 computer terminals through only two IBM 7090 mainframe computers (SABRE 2017). The CRS was exemplar at the time of what were called, "third generation" digital computers and operating systems, which included methods of data and memory management that would form the basis of large-scale commercial computers later incorporated across the tourism and hospitality sector (Denning 1971; Benckendorff et al. 2019, p. 10). Other airlines shortly followed suit to develop their own CRSs, such as PanAm's *PANAMAC* and, later, non-US and multinational and regional developments such as *Galileo* and *Amadeus* in Europe and the *Abacus* CRS serving the Asian and Pacific region. These new national and regional variations were driven by desires to "secure control of the distribution of travel and related tourism products in their own markets" (Truitt et al. 1991).

Beyond operation by the airlines themselves, the first CRS terminals were launched in travel agencies in 1976 (Naqvi and Jia 2014). Over the years, CRSs vastly expanded to include tourism products and services by other vendors, such as hotel rooms, rental cars, train rides, package tours and excursions. While initially only available to the vendors of these products and services, the CRS developed into what is now known as the "global distribution system" (GDS). This computerized networked system came to provide real-time information to travel agents and other resellers on the availability of tourism products and services, who received commissions when selling them. Hoteliers, rental company cars and other commercial entities would connect their inventory of seats, and rooms to one or multiple GDSs (Buhalis 2019).

With the arrival of Personal Computers (PCs), commercial Internet and the World Wide Web through the 1980s and 1990s, networked computerized devices and software became available to the broader public. These first versions of the Web, often termed Web 1.0, allowed users to read content on static websites that were usually produced by Web Coders and Designers (Beer and Burrow 2007). Profiting from this new computerized connection to potential buyers of tourism products and services, Online Travel Agents (OTA) such as Expedia started to emerge through the late 1990s and early 2000s. They have made use of GDSs to provide people (with a PC and internet connection) direct access to expansive databases of available flights, hotel rooms and other tourism-related products through their websites (Buhalis and Law 2008). These websites apply ranking *algorithms* (software) which in technical terms are the instructions with which input data can be transformed in desirable output data (Kitchin 2017). OTAs' algorithms execute travellers' search queries (data inputs) and provide recommendations on what flights, hotel rooms and other products and services to book (data outputs) based on the relevancy of the traveller's particular preferences such as dates, price ranges and past search behaviour. As such, OTAs like Expedia, Booking.com and Skyscanner have come to function as intermediaries between airlines, hotels and other providers, and their end-users (customers) without the necessary need for in-person communication. So-called "Channel Manager" software allows providers and vendors of tourism products and services to automate their inventory across various "channels" (OTAs). These Channel Managers automate availability, rates and reservations in near real-time, reducing the probability of tourism products and services being under- or overbooked. Given that many prospective travellers started to book their trips online, these automized applications had a major impact on the traditional role that brick-and-mortar travel agencies and human agents used to play in the provision of information and booking processes (Novak and Schwabe 2009).

The "locus of control" in the creation and uptake of information in relation to tourism products and services further changed with the emergence of Web 2.0 and the implementation and widespread use of social networking technologies since the early 2000s. Web 2.0 and social networking technologies have allowed for the dynamic production and distribution of online content by anyone with a web enabled interface (Beer and Burrow 2007). For the tourism industry, the onset of Web 2.0 meant that the production and distribution of web content about tourism products and services increasingly became the expression of the interaction and participation of end-users, such as tourism consumers (Munar 2011). A key example in this respect is Tripadvisor, a travel review site founded in 2000 which relies on the voluntary contribution of content by millions of travellers. Here, people share their experiences of tourism-related products, services and entire tourist destinations with other users in the form of ratings and written statements. These forms of assessment aim to inform the choices of other travellers who have intentions to make similar trips. To make the site profitable, Tripadvisor started to advertise online banners and later provided weblinks to tourism providers' own online booking applications as well as those of OTAs. Importantly, the information that people provide and access on myriad websites like Tripadvisor is also extracted and appropriated by the companies that design and operate these websites. Consumer data—whether voluntarily contributed or not—have become one of the most prized commodities in the digital capitalist economy, as value is derived from these data in a variety of ways (Mejias and Couldry 2019; Sadowski 2019). Data can be used to profile and target people, for example, through personal advertising or by charging customers higher prices for tourism products and serviced based on their characteristics and previous search entries. Data has also been used to optimize computerized systems and to make them more productive and profitable. For example, by calculating the estimated weight of people who will board a plane—based on available data about their age or other factors—airline companies aim to use aviation fuel more sparsely and as efficiently as possible (Melis et al. 2019). In yet other ways, data has been used to control processes and people, for example, to monitor, influence and anticipate the movement of travellers through border check points (Amoore 2011). Arguably, tourism's dependency on computerized devices and data has by now become so vast that without these devices and data, the movement of people between spaces and places would likely fail (Kitchin and Dodge 2014).

PLATFORMS AND MARKETS

This brings us to the relatively recent emergence and discursive construct of the "platform", a term that has meaning in different semantic areas, but in its association with online content-hosting intermediaries has come into wider circulation since the mid-2000s (Gillespie 2010; Helmond 2015). Like their impact on other economic sectors, platforms have vastly changed the landscape of tourism over the span of a decade. The computational meaning of a "platform" is "a programmable infrastructure upon which other software can be built and run" (Gillespie 2017) but in public discourse the term "platform" has increasingly been used to describe companies that offer Web 2.0 services, or, networking technologies, that afford their users the opportunity to communicate, interact or sell (Gillespie 2010, pp. 349–351). As such, platforms are commonly seen as intermediaries between supply and demand of a product or a service, between individual users of the web. Deliveroo, for example, is a food-delivery platform that connects riders, restaurants and customers, whereas Facebook is a platform that predominantly provides spaces of connection and socialization for individuals, commercial actors, governmental entities and many others. When understood as a business, a platform's role is to bring together sets of users to configure a so-called *two-sided* or *multi-sided* market (Helmond 2015). Markets that bring together accommodation- and hospitality providers and consumers—operated by platforms like *CouchSurfing* and *Airbnb*—rely on software-enabled digital infrastructures. Technically speaking, anybody who has a property or room available, or who wishes to stay overnight at somebody's home, can become a part of this market. Antecedent operators of multi-sided markets include dating clubs, which establish connections between two or more people, or payment service providers such as VISA or Mastercard, which bring together the market of consumers wishing to pay for something, the producers willing to sell, and finally banks (Barns 2020).

Two-sided and multi-sided tourism markets are different from traditional business-consumer markets. In traditional business-consumer markets, commercial entities such as hotels and motels can sell their products and services directly to their customers and (usually) own the properties and means of production to make profit. Platform companies like *Airbnb*, on the contrary, operate on so-called "lean business models" (Srnicek 2016). They do not own any of the properties that are advertised on their websites and applications, and they do not take any responsibility or pay for the household labour and hospitality that their users have to provide. Instead, these platforms collect rent through user fees, and they rely on (user) data as a means of profit generation and competitive advantage. These platforms are embedded into wider processes of capitalization and ascribe to what has been termed "platform capitalism" (Sadowski 2019; Srnicek 2016; van Dijck et al. 2018; van Doorn 2020). Citing Nick Srnicek (2016), Sarah Barns (2020, p. 102) argues that: "the key advantage of the platform business model over others in an era of big data value chains is the capacity for the platform to occupy a privileged position as intermediary and therefore 'governor' of digital exchange and the data that result as a consequence".

Arguably, platforms only become attractive to prospective users when there is a critical mass of providers and consumers, or, supply and demand available on the platform (Frenken et al. 2020). Since attracting these users is not an easy endeavour, many platform enterprises, therefore, tend to offer, at least initially, the use of their digital infrastructure for free. It has been argued that "once a critical mass of providers and users are active on a platform, more participants will be drawn to the platform so as to profit from the two-sided network effects", effectively enabling platform enterprises to monopolize the markets that they enable (ibid., p. 403).

Despite their apparent differences, business-consumer markets and two- or multi-sided markets do not necessarily exclude each other. Over the years traditional accommodation providers like hotels and motels have increasingly relied on digital platforms like Booking.com and *Airbnb* as intermediaries between them and their customers, aside from operating their own booking sites. In doing so, hotels and other commercial entities have paid advertising fees and commissions to these platforms for having their products and services featured on and booked through these platforms. Although these digital forms of intermediation might be considered relatively new, brick-and-mortar travel agencies have of course fulfilled the function of intermediaries between providers and customers for decades, receiving commission on bookings for myriad tourism products and services, including accommodation rentals.

Anne Helmond (2015, p. 4) argues that for a website (or, software) to be termed a platform, it "needs to provide an interface that allows for its (re)programming: an API" (Application Programming Interface). APIs allow for data exchange between applications, evidenced in the example of dating app *Tinder*, which has been built on top of the *Facebook* platform. *Twitter* allows "users to login with Facebook and uses Facebook data such as 'likes' and shared friends to match potential partners" (ibid.). Similarly, *Airbnb*'s introduction of "*Airbnb* Social Connections" in 2011 have allowed *Airbnb* users to log in to the *Airbnb* website with their *Facebook* accounts, which could make visible which *Facebook* friends had stayed with certain hosts or were friends with certain hosts (Presenza et al. 2021). At the same time, selected external third parties and developers have been able to make use of *Airbnb*'s API to onboard their own inventory of properties and listings (Airbnb 2021a, b). Accordingly, platforms have "made their content and functionality available as part of a business strategy in which third parties can add value to a platform by building new services on top of it" (Helmond 2015, p. 4). In this way, platform companies can expand their presence in other (social) spaces, under certain conditions. What makes platforms "platforms" are that thus they are "programmable".

Platforms and the "Sharing Economy"

Platforms that facilitate interactions between people have often been associated with the "sharing economy", in particular, platforms that enable the sharing or renting (out) of resources with "excess capacity". The commonly held argument is that excess capacity of a consumer good is present when the owner of such a resource does not utilize or consume it all the time. Many if not most consumer goods have been understood to potentially have excess capacity, including cars, boats, rooms, houses, clothing, books, toys, appliances, tools, furniture, computers and even pets. In the case of *Couchsurfing* and *Airbnb*, users of these platforms may share or rent (out) "underutilized" rooms or entire homes. In the case of *BlaBlaCar*, a major ride-sharing platform in Europe, people may temporarily share "underutilized" seats in their car to get from A to B, dividing the cost of transport between them. This temporary access to other people's goods and services, possibly for money, has been constitutive of what has been termed the "sharing economy", an economic configuration that has supposedly instigated a shift away from the traditional top-down exchange models and ownership models (Frenken and Schor 2017, pp. 4–5). In public discourse the term "sharing economy" became particularly popular around 2010 when a best-selling book *Collaborative Consumption* by Rachel Botsman and Roo Rogers (2010) proclaimed that the new sharing economy would have many expected sustainability impacts. The idea that was promoted in this book was that consumers would get cheap access to goods by renting (out) or lending them to/from others: people would consume collaboratively. By consuming collaboratively fewer new goods would have to be produced and wasteful overconsumption could be countered, at least, that was the hypothesis. This additionally would come with environmental benefits, specifically in the case of car sharing. As cars stand idle most of the time, any type of sharing scheme that made cars accessible to non-owners would reduce the number of cars being produced and would promote higher utility of a car (Frenken and Schor 2017). The sharing economy would also come with positive social externalities: because platforms are built on networking technologies, people who did not know each other from before could get in touch, and would be able to share, thereby extending an existing practice to a larger social scale. Some have claimed that the sharing economy was/is a new phenomenon, but put in economic-historic perspective, the practice of "sharing" is nothing essentially new or innovative (Belk 2014). People have always shared for functional reasons such as survival, but also out of altruistic motivations, based on social and cultural norms. Sharing in everyday life reproduces social relations and is seen to strengthen cultural practices; it is not just a relic of pre-modern societies; it is a practice we engage in for survival almost every day. Food-sharing, for example, is an important social and cultural tradition that not only secures sustenance but solidifies social bonds and community cohesion (Davies 2020, p. 205). Collaborative forms of consumption and sharing economies that operate outside of more traditional, capitalistic modes of exchange have, in fact, existed for millennia.

Critical analyses of the "sharing economy" have highlighted the inherent plurality of economic forms that are seemingly unproblematically clustered together in the same economic and discursive space (Ince and Hall 2017, p. 3). While some platforms associated with the sharing economy surely facilitate new collaborative and communal forms of economic activity, other platforms mirror "traditional" capitalist relations of production, which also "impact negatively on their clients, workers and broader economic environments" (ibid.). Koen Frenken and Juliet Schor (2017) provide a conceptual framework for understanding the different types of platforms that constitute the sharing economy, which adhere to the following characteristics: (1) they are based on consumer-to-consumer (c2c) interactions, and (2) they allow for *temporary* access to, (3) physical goods. Abstractedly speaking, these characteristics have been applicable to platforms such as *Couchsurfing* and *Airbnb* since they facilitate a market between individual hosts and guests who temporarily share or rent (out) rooms or entire properties. Yet, for-profit platforms like *Airbnb* also increasingly involve transactions between businesses and consumers rather than between consumers (c2c) alone and rooms and entire properties can be rented out on a permanent basis. Thus, in line with Ince and Hall (2017), the optimism surrounding the "sharing economy" and its apparent unproblematic association with for-profit short-term rental platforms, should be carefully interrogated rather than taken for granted, as will be exemplified in Chapter 3.

Frenken and Schor also branch out three different types of platforms that do not fall under the sharing economy banner since they exclude one of several of these characteristics. Firstly, the on-demand or "gig" economy, in which platforms bring together the supply of, and demand for, labour rather than goods. Examples in tourism are *Uber* and *Hlprs*, platforms which connect people who offer/require taxi services or domestic services, respectively. The software of these platform companies usually incorporates fee-systems, determining the wages of platform workers based on factors such as labour availability, distance and the time that it takes to complete a task. These platforms' price-setting algorithms also determine the market in that they decide who gets connected to who and set the conditions of labour. On-demand platforms could therefore be conceived of as platform labour intermediaries that operate as new players in a dynamic temporary staffing industry (van Doorn 2017). Their software works to "optimize labor's flexibility, scalability, tractability, and its fragmentation" (ibid., p. 901). Another economy that falls outside of the sharing economy label is the Product-Service Economy. Here, individuals can rent goods from a company rather than from another consumer. For example, car-rental company *Hertz* gives people temporary access to cars, while bike-rental company *BITS* gives people temporary access to bikes. Rather than being an economy between individual consumers, the Product-Service Economy is a market between businesses and consumers. Finally, Frenken and Schor discuss the Second-Hand- and Gift Economy. In these markets people sell or gift each other second-hand or unused goods, changing the ownership status of physical goods and providing permanent access. For example, the British platform *Olio* enables people living close to each other to gift each other unused food or medicine. Italian platform *Subito* enables people to sell or gift second-hand goods such as furniture and cars. However, despite practices of "sharing" being evident in these economies, they do not fall under Frenken and Schor's proposed framework of the sharing economy since consumers grant each other *permanent* access rather than temporary access to their goods.

What is clear is that the term "sharing economy" has worked as a smokescreen that obscures the complex and sometimes opposing economic philosophies that sustain it. Moreover, the idea of sharing also reveals little about the kind of public space and infrastructure that many of these platforms rely on and how that reliance may produce negative outcomes and unsettle certain public interests (van Dijck et al. 2018). While platforms tend to have different organizational forms and impacts on society, what commonly unites them is their role in facilitating exchange between several different user groups. Moreover, they are enabled by digital infrastructures developed by companies, corporations or organizations, who in one way or another, extract value out of these exchanges and the data that users produce in the process. In this respect, Sarah Barns draws attention to the critical role that is played by data architecture in the "sharing economy". The protocols and standards that underpin platforms ensure that platform companies "continuously learn from, extract, and commodify the informational outputs of all 'sharing' they facilitate" (2020, p. 15). Far from being neutral intermediaries between those who engage in economic transactions, platform companies have the potential to govern users' behaviour and transactions in specific directions. In the context of tourism and hospitality, this argument will be exemplified in Chapter 5 of this book.

HISTORICAL CONTEXT OF HOSPITALITY NETWORKS AND PLATFORMS

Although digitized short-term rental- and hospitality networks like *Couchsurfing* and *Airbnb* have gained tremendous popularity since the 2000s, "home-sharing" as a source of supplemental income—also termed "lodging"—has a long history (see Goyette 2021; Dallen and Teye 2009). Moreover, earlier forms of institutionalized "analogue" accommodation sharing and hospitality networks pre-date the era of commercial internet era. These networks relied on paper lists of data (spreadsheets), printed address guides, telephone communication and regular post to connect hosts and guests in the network. Many of these pre-internet networks still exist today although many of them now operate their services and/or databases through (static) websites to facilitate their sharing economies between hosts and guests. Interestingly, many of these networks have created successful niches by targeting individuals along specific social categories such as age, gender, occupation, sexuality, nationality and religion. Others have branched out by catering to people with specific hobbies; speaking certain languages or travelling to certain regions. For example, *Servas*—a non-profit, non-governmental, international network of hosts and travellers—was initiated in 1949 to promote peace after World War II. Other examples are *Hospitality Exchange* or *Hospex.org*, a hospitality network established in the early 1960s aimed at bringing travellers together; *Friendship Force International* (est. 1977), which, as its name suggests, aspires to enforce friendship through travel; *Pasporta Servo* (est. 1979) a network specifically for Esperantists or coming-to-be Esperantists; *Women Welcome Women* (est. 1984) a network encouraging women befriending women through hospitality; *LGHEI* (est. 1991) a hospitality network for lesbian, gay, bisexual and transgender travellers; *The Affordable Travel Club* (1992) a hospitality and accommodation exchange network for senior and baby boomer' travellers and *Vrienden op de Fiets* (transl.: *Friends on Bikes*, est. 1984) or *WarmShowers* (est. 1993), which facilitate hospitality for touring cyclists.

These networks, followed by their digital descendants, are composite and heterogeneous in terms of travel objectives, philosophies and ideologies, but they broadly speak to collaborative or relational forms of travel. Their purpose is to connect different user groups in their exchange of services, and in the process, generate value. The advent of personal computers, the Internet, social networking sites and web content management systems in the early and mid-1990s, inspired many of these networks to profile themselves online. It also afforded many new and *natively* digital initiatives to be developed. Non-monetary and non-profit hospitality and accommodation exchange platforms like *Belodged.com* (1999), *Hospitality Club* (2000), *Couchsurfing* (2003), *Bewelcome* (2007) and *Globalfreeloaders* (2010) began to facilitate the exchange of non-commercial accommodation worldwide, relying on the reciprocity and trust among its users to advance exchanges. Cutting out traditional commercial intermediaries, these platforms mostly leaned on ideologies pertaining to "moral travel", anti-consumerist culture and anti-mass-and/or organized tourism (Picard and Buchberger 2013). Among non-monetary platforms are numerous globally organized volunteering platforms such as *WWOOF* and *HelpX*, which provide room and board in exchange for labour. Other platforms have experimented with different remuneration systems that are non-monetary or combine monetary with non-monetary remuneration. The French start-up *GuestToGuest* (now *HomeExchange*) for example, operates by a point system as a form of credit. Other platforms such as *Staydu* (2011) and *SabbaticalHomes* (2000) leave the choice between exchanging, sharing, renting or "sitting" to its users, and experiment with multiple forms of remuneration.

However, platforms that operate monetary exchange systems have gradually crowded out non-commercial, more egalitarian and democratic forms of hospitality and accommodation exchange. Examples are *VRBO* (1995), *Homeaway* (2005), *FlipKey* (2006), *Tripping* (2010), *9flats* (2010), *Wimdu* (2011), *Travelmob* (2012), *BeMate.com* (2014) and finally giant *Airbnb* (2008), which has since its inception acquired competing platforms such as *Accoleo* (in 2011), *Luxury Retreats* (in 2016), *Trip4Real* (in 2016), *Accomable* (in 2017) and *Crashpadder* (in 2012). These platforms collect booking commissions from their users, be it hosts, guests or both.

A Brief History of *Airbnb*

In the following chapters of this book, *Airbnb* will serve as a case study to discuss a platform's impacts on cities, neighbourhoods and everyday life at home. To offer some contextualization, the following paragraphs first provide an outline of the platform's relatively brief history, its main objectives and a description of its digital architecture. *Airbnb* was co-founded in 2008 by industrial designers Joe Gebbia, Brian Chesky and technical architect Nathan Blecharczyk, who, at the time, were all based in San Francisco (California, USA). Their idea for the platform sprouted at the advent of an International Design Convention in 2007. As all hotels seemed to have sold out in the run-up to the conference, they decided to set up a website through which they advertised three airbeds in their apartment, which conference attendants could book. The money that they received would help cover the rent that they were struggling to pay. Their airbeds were booked, stimulating further aspirations among the three men to create a platform (initially www.airbedandbreakfast.com), which would facilitate the short-term rental of beds in people's homes. It would take until 2009 for the initiative to receive its first round of funding, for the company to be renamed *Airbnb*, and for the platform to start facilitating people to rent (out) entire properties (Airbnb 2021a, b).

Since its inception, *Airbnb* has frequently described itself as an intermediary between travellers (or "guests") and hosts who rent out beds, rooms or entire properties. Guests may book a room *on-site*, which entails renting a (shared) bedroom within a host's home, with the host typically physically present throughout the stay. Alternatively, a guest may rent a host's entire accommodation without the host being physically present during a guest's stay, also termed "remote hospitality" (Ikkala and Lampinen 2015, p. 1036). Through various global brand campaigns, *Airbnb* has frequently critiqued the standardized tourist offerings that "modern tourism" has to offer and has claimed that engaging in mundane practices in other people's homes and neighbourhoods is a better alternative than mass-produced and impersonal travel experiences that are promoted elsewhere. Using *Airbnb*, travellers can live in other people's homes and supposedly experience what it means to "live like a local". However, guests may still receive similar services from their hosts as commonly provided in commercial tourist accommodations such as a check-in, room cleaning, breakfasts and the provision of information before and during the stay.

In the span of a decade, *Airbnb* has become one of the biggest platforms of scale and market capitalization. It has assisted in 1 billion total guest arrivals and it advertised over 5.6 million listings in 2021 alone (Airbnb 2021a, b). The platform is currently viewed as one of the most valuable hospitality companies worldwide, a status often attributed to the billions of US$ it has received in venture capital investment. Despite the gloomy prospects that have been held for the tourism- and hospitality sector since the outbreak of the COVID-19 (coronavirus) pandemic, *Airbnb*'s initial public offering on the stock market in December 2020 secured a valuation of close to US$100 billion, (Allyn and Schneider 2020). Its success has often been ascribed to its insistent deployment of PR, sales and marketing strategies to attract new users, with expenditures estimated up to 1 billion US$ a year (in 2018 and 2019). Additionally, *Airbnb* has been known to engage persistently in lobbying practices to influence lawmakers and massage local and national government relations with the purpose of transforming the regulatory landscape to its advantage (van Doorn 2020; Adamiak 2021). Since its inception, the company has been headquartered in San Francisco and has opened 23 office locations across 15 countries up until today. Local *Airbnb* offices are generally made up of teams of software engineers, data scientists, customer experience specialists, host and community specialists, designers, brand managers and business- and product managers.

Over the years, *Airbnb* has undergone significant changes in terms of design and functionality. It has reinvented itself by exploring new ways of capitalizing on the everyday lives and intimate spatialities of its users by altering its products and services. The platform enterprise remains predominantly known for its role in providing a marketplace of "short-term rentals": it brings together providers and consumers of accommodation, who usually rent these (out) for periods of days, weeks or months. However, since the outbreak of the COVID-19 (coronavirus) pandemic, *Airbnb* has started to nudge its hosts into providing mid-term and long-term stays (Roelofsen and Minca 2021). The pandemic has provoked a large increase in flexible, home- and desk-based work arrangements. These shifts to "remote work" or "tele-work" have apparently driven a specific and privileged group of travellers (sometimes referred to "digital nomads") who look for mid- to long-term accommodation in destinations away from "home" where they can continue carrying out their work while enjoying some benefits they would have while holidaying. *Airbnb* has thus urged its hosts to consider lengthening the period of their bookings, as well as to offer "self-check-in" modalities to their guests, in this way minimizing in-person contacts (Roelofsen and Minca 2021). This recent turn to mid- and long-term stays in private accommodation on part of *Airbnb* will possibly only come to the advantage of a specific group of hosts, namely, those who possess and are (legally) allowed or able to rent out entire properties for long periods of time. Studies have shown that these hosts are usually already able to capitalize on housing assets in ways that generate a more reliable income stream through the platform, and subsequently, expand the scale and revenues of *Airbnb*'s operations (van Doorn 2020, p. 13; Bosma 2021).

As Sarah Barns (2020, p. 101) has observed, there is a need to stay attentive to these kinds of changes that platforms undergo since those changes are not just "a generalised condition of platform capitalism" but a set of strategic interventions on the part those companies to exert and maintain control over their markets. This can also be observed in other changes made to the platform, some having more success than others. In 2017 *Airbnb* invested 13 million US$ in restaurant reservation app *Resy* and added a *Restaurants* application to its platform, which allowed users to book restaurant tables in their respective destinations. However, today this function can no longer be found on the platform, alluding to unsuccessful venture capital investment strategies which have remained underreported in the media (see also Langley and Leyshon 2017). A more successful development strategy on part of *Airbnb* was its incorporation of the *City Hosts* programme (now called *Experiences*) in 2016, an application that allows guests to book a range of activities with *Airbnb* hosts against a set fee, for example, guided tours, cooking workshops or pottery classes. These expansions have been part of a broader attempt to turn idle resources (kitchens, people's time, etc.) "into maximally productive assets and commodifying latent space in existing places" (Sadowski 2019, p. 450).

HOSTS, GUESTS, AND LABOUR IN THE *AIRBNB* ECONOMY

In the *Airbnb* economy, the categories of "host" and "guest" are interchangeable since those who host on the platform can simultaneously use the platform as a guest and vice versa. Creating and maintaining a hospitable space and a pleasant ambiance, often requires emotional- and affective labour on part of both host *and* guest, especially when they stay together in the same property. Appropriate and hospitable behaviour is further incentivized through the review and rating system incorporated in the platform's architecture and economy. Both hosts and guests review and rate each other after a stay and are held accountable for hospitable behavior, which is assessed according to certain standards attributed to their roles. For example, a guest will be rated for how well they have observed and abided to the house rules while a host will be rated for their timely responses to a guest's request. Chapter 5 will discuss the impact of reviewing and rating on performances and understandings of hospitality.

Airbnb "guests" are usually those who book and pay for the listed accommodation and (in most cases) stay in the booked accommodation, with or without the host(s) present. "Hosts", in the most basic definition of the term, are considered those who advertise beds and/or their homes on the *Airbnb* platform against a set fee. The definition of a "host" in the *Airbnb* economy is a contentious one since those whose profiles are listed on the platform are not necessarily those in charge of any of the hosting activities. Neither do these profiles necessarily represent the person who owns or lives in the properties that are advertised, since they may be managed by third parties like property management agencies who take charge of all hosting duties. In Chapter 3, a separate section on professionalization practices on the *Airbnb* platform will shine more light on this matter. Generally speaking, hosting duties entail administrative duties: ensuring the listing is represented online, updating the listing's availability in the calendar software and carrying out bookkeeping tasks; communicating with *Airbnb* guests before, during, and sometimes after their stay; cleaning the rooms and/or the entire property; receiving the guests and/or ensuring key pickup and drop-off; and offering guidance and support during a guest's stay. In case of "remote hospitality", interactions between hosts and guests are usually less frequent than those who physically share the same space, and hospitality is limited to key exchange and an induction to the accommodation. In this case, most hosts will stay available for support at a distance.

Co-Hosts and Reproductive Labour

In the *Airbnb* economy, it is increasingly common to have third parties involved in the provision of certain aspects of hospitality labour. Hosting duties are frequently delegated after online bookings are made, with or without prior consent of the guest. A host may delegate their hosting activities to enlisted "co-hosts", partners, friends, family members or tenants, if they, for whatever reason, cannot be physically present during the guest's arrival or stay. Additionally, property management agencies such as *Luckey* in Spain and *Airbnbutler* in the Netherlands, can assume responsibility for all hosting activities. These companies form a whole new support economy to "strengthen" and "professionalize" the short-term rental economy are flourishing globally. The services that these companies offer may include answering inquiries, key exchange, hosting guests throughout their stay, writing reviews and assuming cleaning duties. In some cases, communication with a host may remain entirely online and hosts and guests may never meet in person. Directions to the location of the house keys may be communicated through *Airbnb*'s messaging system or by phone, as well as the "house manual" or "house rules", which may come in a paper or digital form. Access to the accommodation may also be enabled through technology, for example, through code-operated door locks. In the latter case, physical contact with the host may be entirely absent.

With *Airbnb*, domestic space and reproductive labour—the mental, manual and emotional labour needed to maintain life, such as housework and care—are (temporarily) monetized. For those hosts who take guests into their homes, added reproductive labour is required in their households, due to the different needs that guests have. For example, rooms must be prepared for use by the guest in ways that often resemble hotel room preparation such as decluttering, (re)making beds with clean bedlinen and engaging in communication with guests during their stay to reassure their needs are met (Roelofsen and Goyette 2022). The divisions of this added labour in *Airbnb*-ed homes often reflect existing gender ideologies within households that consist of both men and women. Women usually carry out manual labour such as cleaning while men maintain the managerial work, such as managing the (financial) administration (ibid.). Such divisions of labour are usually also maintained by hosts who rent out entire homes and are not present during a guest's stay: it is usually men who rent out their properties on *Airbnb* but outsource the manual cleaning labour to a hired help, who are typically women. In Chapter 4, the question of labour and *Airbnb*'s potential to destabilize or transform household relations will be dealt with in more depth.

AIRBNB'S DIGITAL ARCHITECTURE

In terms of digital architecture, the *Airbnb* platform has almost all the building blocks of a traditional Online Travel Agency to collect, store, process and appropriate data. In Chapter 5, I provide a more in-depth analysis of *Airbnb*'s data gathering practices and the implications these practices have for the users that participate in the *Airbnb* economy. The platform incorporates a vast number of applications, or software (or "systems" as *Airbnb* calls them) that are typically used in the hotel and hospitality branch and exist as such since the 1980s.[1] The Property Management System (PMS), for example, is one of *Airbnb*'s main applications that facilitates the type of front office and back-office work that can also be observed in the hotel industry such as revenue and reservation management, guest information processing, room and rate assignment, check-in and check-out management and accounting. In the case of *Airbnb*, this software allows for the reservation and management of properties (bedrooms, homes) and/or touristic experiences enlisted by hosts, and the storing and processing of user data. The PMS and other applications that are incorporated in *Airbnb*'s architecture require data to operate. They also need a constant supply of *new* data to keep the accommodation marketplace that they sustain running. In the hotel sector, PMS systems are usually operated by front- and back-office workers with expert knowledge of booking software and the skills to inform customers about the outputs that these systems produce. In the case of platforms like *Airbnb* and *Couchsurfing*, anyone with a computer and internet connection is expected to understand and operate the software without prior knowledge. Users of the *Airbnb* platform, typically proceed through *Airbnb*'s search engine after entering the main webpage of *Airbnb* as they look for listings and/or touristic experiences in particular locations around the world. *Airbnb*'s search engine is organized around "match-making" algorithms that predict which hosts and guests may have matching expectations and experiences and determines a guest's search results based on a number of factors including, but not limited to, the type of listing (an entire place, a room or shared room), the number of nights they wish to stay, the price, and, not in the least the location. When guests carry out a search command through the platform, the results are always presented by the algorithm in a specific ranking of listings that is "geolocalized and transferred to map-based interfaces" (Celata et al. 2020, p. 131).

The *Airbnb* platform also features a messaging application, a payment application, an invoicing application and a review and rating application, which allows both hosts and guests to rate and review each other after their stay or experience. It also includes a statistics and metrics application that manages and displays the performances of both guests and hosts

---

[1] https://www.airbnb.com/software-partners and https://www.airbnb.com/software-partners/preferred-partners?selectedSoftwareType=PROPERTY_MANAGEMENT_SYSTEM.

(and their listing(s)). Also incorporated is a "pricing tool" that suggests "competitive" overnight rates for their advertised listing based on the availability of and demand for other *Airbnb* property in the area. This tool takes into account that certain nights of the week (e.g. weekends) are in higher demand than others and that seasonality and popular events in the host's area may allow hosts to ask higher overnight rates.

Like Online Travel Agents, *Airbnb* primarily earns its revenues through the fees it charges both hosts and guests who partake in a transaction. Although the fees have changed over the years, today the guest service fee amounts to 14% of the guest's reservation subtotal (before fees and taxes). The host service fee is typically 3%, which is deducted from the amount that the host would receive from the guest. Hosts who offer "*Airbnb* experiences" are charged a 20% service fee, which is calculated from the price that hosts set for their experience. Depending on the hosts' local laws of the jurisdiction, VAT is charged on top of the host and guest service fee.

POSITIONING *AIRBNB*

As argued above, *Airbnb* as an *economic activity*—renting out private space and providing hospitality—and as a *business model*—deriving commissions from a digitally enabled network of buyers and suppliers of accommodation—is nothing essentially new. Myriad globally networked hospitality initiatives preceded the advent of *Airbnb* and the business of lodging has existed for centuries. Moreover, numerous software applications (or systems) that *Airbnb* has incorporated in its digital architecture, already existed in different forms for decades and were extensively used in the hotel and hospitality industry. However innovative *Airbnb* is propagated to be, its strategies for capital accumulation and unbound growth reflect those that have driven the innovation and development of the first Computer Reservation Systems since the 1950s and the first Online Travel Agencies since the late 90s and early 2000s. The platform's purpose is to profit from an ever-expanding network of data that drives its economy. Like the early CRSs and OTAs, *Airbnb* automatically processes reservations and sales as *efficiently* as possible, while deriving commissions from both buyers (guests) and sellers (hosts) in their transactions. Like many of the aforementioned digital innovations that preceded *Airbnb* in the last 7 decades, the platform enterprise is driven by capitalist modes of economic organization. *Airbnb* encourages current and future users to rent out properties and material resources—otherwise only privately used—as "a viable mode not only of managing and distributing those resources but also extracting profit from them" (Ince and Hall 2017, p. 3). These "resources" (homes and rooms) and the labour and time needed to prepare them for guests to use, have supposedly been made "productive" by the platform; homes and rooms are turned into productive assets that generate rents (Sadowski 2019). By (re-)developing web applications that could be easily accessed by lay users, *Airbnb* has sought to secure control of the already existing short-term (or vacation) rental market, mirroring monopolistic tendencies observed in CRSs that developed systems like the first SABRE CRS in the 1960s. Together with other accommodation platforms such as Booking.com, Agoda and Expedia, *Airbnb* now dominates the global "short-term rental" market (Adamiak 2019) ensured, in part, by its acquisition of various competitor platforms.

In yet other ways, the *Airbnb* economy reproduces the socio-economic and racial inequalities that have pervaded the tourism and hospitality industry long before the advent of platforms (Benjamin and Dillette 2021). What has emerged from numerous studies is that *Airbnb* mainly benefits already privileged, asset-heavy, middle-class and white members of society (Bosma 2021; Schor 2017; Roelofsen 2018) rather than the "ordinary residents" who, according to *Airbnb*, try to "make ends meet" (Airbnb 2020; Goyette 2021). Studies have evidenced the emergence of forms of implicit and explicit racial discrimination and social exclusion on short-term rental platforms like *Airbnb*. Despite recent anti-discrimination policies, platform markets lack regulatory monitoring and protections against discrimination (Piracha et al. 2019). Their marketplaces require people to visually and textually profile themselves while at the same time allowing people to refrain from or refuse exchanging with people of certain racial backgrounds (Edelmann et al. 2017). Studies have shown that bookings by *Airbnb* guests with distinctively African American names tend to be accepted less than guests with white American names (Piracha et al. 2019). Meanwhile, in response to discrimination and exclusion, alternative platforms have been exclusively conceived for those in search for safe(r) and more inclusive economic spaces, although their relative success has often been short-lived. *Muzbnb* or *Innclusive*, for example, were platforms that operated exclusively for Muslim travellers and travellers of colour but have seized to exist.

What is perhaps "new" about *Airbnb*, is that it has managed to gain a dominant position in the short-term rental and lodging sector while largely flaunting, objecting to, and co-shaping the existing rules and regulations that apply to rental housing and the tourism accommodation sector. Although it has enthusiastically promoted that anyone can become a host or guest on the platform, it has taken little responsibility to ensure that the practice of renting out residential housing for touristic, short-term purposes is in line with existing tourism and local housing laws. Instead, the platform enterprise has deployed a range of strategies—through lobbying and marketing campaigns, but also through litigation—to convince local governments and society of its positive impacts in cities and to influence the terms of current and future policy on short-term renting to favour its purposes (van Doorn 2020). It has also mobilized its own users, hosts in particular, to campaign and vouch for *Airbnb*'s role as an essential urban asset. Niels van Doorn therefore has framed *Airbnb* as an *urban institution* which has become a powerful social player in the governance of many cities (ibid.). In the following chapters I detail the consequences of *Airbnb*'s approach to business, governments and society, detailing its socio-spatial impacts on cities, housing, neighbourhoods and the everyday lives of its users.

References

Adamiak, Czesław. 2019. "Current State and Development of Airbnb Accommodation Offer in 167 Countries." *Current Issues in Tourism*, December, 1–19. https://doi.org/10.1080/13683500.2019.1696758.

Adamiak, Czesław. 2021. "Changes in the Global Airbnb Offer During the COVID-19 Pandemic." *Oikonomics* May (15). https://doi.org/10.7238/o.n15.2107.

Airbnb. 2021a. *Fast Facts*. https://press.atairbnb.com/fast-facts/.

Airbnb. 2021b. *Connect to Our API. Connect to Millions of Travelers on Airbnb*. https://www.airbnb.com/partner.

Airbnb. 2020. "Hosting Financials: Bringing Home Ownership within Reach." 2020. https://www.airbnb.com/d/homeownership.

Allyn, Bobby, and Avie Schneider. 2020. "Airbnb Now A $100 Billion Company After Stock Market Debut Sees Stock Price Double." *NPR*, December 11, 2020. https://text.npr.org/944931270.

Amoore, Louise. 2011. "Data Derivatives: On the Emergence of a Security Risk Calculus for Our Times." *Theory, Culture & Society* 28 (6): 24–43. https://doi.org/10.1177/0263276411417430.

Barns, Sarah. 2020. *Platform Urbanism*. Singapore: Springer Singapore. https://doi.org/10.1007/978-981-32-9725-8.

Beer, David, and Roger Burrows. 2007. "Sociology and, of and in Web 2.0: Some Initial Considerations." *Sociological Research Online* 12 (5). http://www.socresonline.org.uk/12/5/17.html.

Belk, Russell. 2014. "You Are What You Can Access: Sharing and Collaborative Consumption Online." *Journal of Business Research* 67 (8): 1595–1600. https://doi.org/10.1016/j.jbusres.2013.10.001.

Benckendorff, Pierre J., Zheng Xiang, and Pauline J. Sheldon. 2019. *Tourism Information Technology*. Oxfordshire, UK: Cabi.

Benjamin, Stefanie, and Alana K. Dillette. 2021. "Black Travel Movement: Systemic Racism Informing Tourism." *Annals of Tourism Research* 88 (May): 103169. https://doi.org/10.1016/j.annals.2021.103169.

Bosma, Jelke R. 2021. "Platformed Professionalization: Labor, Assets, and Earning a Livelihood Through Airbnb." *Environment and Planning A: Economy and Space*, January, 0308518X211063492. https://doi.org/10.1177/0308518X211063492.

Botsman, Rachel, and Roo Rogers. 2010. *What's Mine Is Yours: The Rise of Collaborative Consumption*. New York: Harper Collins.

Buhalis, Dimitrios. 2019. "Technology in Tourism-from Information Communication Technologies to ETourism and Smart Tourism Towards Ambient Intelligence Tourism: A Perspective Article." *Tourism Review* 75 (1): 267–72. https://doi.org/10.1108/TR-06-2019-0258.

Buhalis, Dimitrios, and Rob Law. 2008. "Progress in Information Technology and Tourism Management: 20 Years on and 10 Years After the Internet—The State of ETourism Research." *Tourism Management* 29 (4): 609–23. https://doi.org/10.1016/j.tourman.2008.01.005.

Celata, Filippo, Cristina Capineri, and Antonello Romano. 2020. "A Room with a (Re)View. Short-Term Rentals, Digital Reputation and the Uneven Spatiality of Platform-Mediated Tourism." *Geoforum* 112: 129–38. https://doi.org/10.1016/j.geoforum.2020.04.007.

Davies, Anna R. 2020. "Food Sharing." In *Routledge Handbook of Sustainable and Regenerative Food Systems*, edited by J. Duncan, M. Carolan, and J.S.C. Wiskerke, 204–17. London: Routledge.

Denning, Peter J. 1971. "Third Generation Computer Systems." *ACM Computing Surveys (CSUR)* 3 (4): 175–216.

Dijck, José Van, Thomas Poell, and Martijn De Waal. 2018. *The Platform Society: Public Values in a Connective World*. Oxford University Press.

Doorn, Niels van. 2017. "Platform Labor: On the Gendered and Racialized Exploitation of Low-Income Service Work in the 'on-Demand' Economy." *Information, Communication & Society* 20 (6): 898–914. https://doi.org/10.1080/1369118X.2017.1294194.

Doorn, Niels van. 2020. "A New Institution on the Block: On Platform Urbanism and Airbnb Citizenship." *New Media & Society* 22 (10): 1808–26. https://doi.org/10.1177/1461444819884377.

Edelman, Benjamin, Michael Luca, and Dan Svirsky. 2017. "Racial Discrimination in the Sharing Economy: Evidence from a Field Experiment." *American Economic Journal: Applied Economics* 9 (2): 1–22. https://doi.org/10.1257/app.20160213.

Frenken, Koen, Arnoud van Waes, Peter Pelzer, Magda Smink, and Rinie van Est. 2020. "Safeguarding Public Interests in the Platform Economy." *Policy & Internet* 12 (3): 400–425. https://doi.org/10.1002/poi3.217.

Frenken, Koen, and Juliet Schor. 2017. "Putting the Sharing Economy into Perspective." *Environmental Innovation and Societal Transitions* 23: 3–10. https://doi.org/10.1016/j.eist.2017.01.003.

Gillespie, Tarleton. 2010. "The Politics of 'Platforms.'" *New Media and Society* 12 (3): 347–64. https://doi.org/10.1177/1461444809342738.

Gillespie, Tarleton. 2017. "The Platform Metaphor, Revisited." Culture Digitally. 2017. http://culturedigitally.org/2017/08/platform-metaphor/.

Goyette, Kiley. 2021. "'Making Ends Meet' by Renting Homes to Strangers." *City* 25 (3–4): 332–354. https://doi.org/10.1080/13604813.2021.1935777.

Helmond, Anne. 2015. "The Platformization of the Web: Making Web Data Platform Ready." *Social Media + Society* 1 (2): 2056305115603080. https://doi.org/10.1177/2056305115603080.

Ikkala, Tapio, and Airi Lampinen. 2015. "Monetizing Network Hospitality." In *Proceedings of the 18th ACM Conference on Computer Supported Cooperative Work & Social Computing*, 1033–44. New York, NY, USA: ACM. https://doi.org/10.1145/2675133.2675274.

Ince, Anthony, and Sarah Marie Hall (eds.). 2017. *Sharing Economies in Times of Crisis: Practices, Politics and Possibilities.* New York: Routledge.

Kitchin, Rob. 2017. "Thinking Critically About and Researching Algorithms." *Information Communication and Society* 20 (1): 14–29. https://doi.org/10.1080/1369118X.2016.1154087.

Kitchin, Rob, and Martin Dodge. 2014. *Code/Space: Software and Everyday Life.* MIT Press.

Langley, Paul, and Andrew, Leyshon. 2017. Platform Capitalism: The Intermediation and Capitalization of Digital Economic Circulation. *Finance and Society* 3 (1): 11–31. https://doi.org/10.2218/finsoc.v3i1.1936.

María Munar, Ana. 2011. "Tourist-Created Content: Rethinking Destination Branding." Edited by Rich Harrill and Leonardo Dioko. *International Journal of Culture, Tourism and Hospitality Research* 5 (3): 291–305. https://doi.org/10.1108/17506181111156989.

Mejias, Ulises A., and Nick Couldry. 2019. "Datafication." *Internet Policy Review* 8 (4). https://doi.org/10.14763/2019.4.1428.

Melis, Damien J., Jose M. Silva, Miguel A. Silvestre, and Richard Yeun. 2019. "The Effects of Changing Passenger Weight on Aircraft Flight Performance." *Journal of Transport & Health* 13: 41–62. https://doi.org/10.1016/j.jth.2019.03.003.

Naqvi, Masood A., and Hongyan Jia. 2014. "Computer Reservation System, Tourism BT." In *Encyclopedia of Tourism*, edited by Jafar Jafari and Honggen Xiao, 1–3. Cham: Springer International Publishing. https://doi.org/10.1007/978-3-319-01669-6_510-1.

Novak, Jasminko, and Gerhard Schwabe. 2009. "Designing for Reintermediation in the Brick-and-Mortar World: Towards the Travel Agency of the Future." *Electronic Markets* 19: 15–29. https://doi.org/10.1007/s12525-009-0003-5.

Oxley, David, and Chaitan Jain. 2015. "Global Air Passenger Markets: Riding out Periods of Turbulence." *IATA The Travel & Tourism Competitiveness Report*, 59–61.

Picard, David, and Sonja Buchberger (eds.). 2013. *Couchsurfing Cosmopolitanisms. Can Tourism Make a Better World?* Transcript Verlag.

Piracha, A., Sharples, R., Forrest, J., & Dunn, K. 2019. "Racism in the Sharing Economy: Regulatory Challenges in a Neo-liberal Cyber World." *Geoforum*, 98, 144–152.

Presenza, Angelo, Umberto Panniello, and Antonio Messeni Petruzzelli. 2021. "Tourism Multi-Sided Platforms and the Social Innovation Trajectory: The Case of Airbnb." *Creativity and Innovation Management* 30 (1): 47–62. https://doi.org/10.1111/caim.12394.

Roelofsen, Maartje. 2018. "Performing 'Home' in the Sharing Economies of Tourism: The Airbnb Experience in Sofia Bulgaria." *Fennia—International Journal of Geography* 196 (1): 24–42. https://doi.org/10.11143/fennia.66259.

Roelofsen, Maartje, and Claudio Minca. 2021. "Sanitised Homes and Healthy Bodies: Reflections on Airbnb's Response to the Pandemic." *Oikonomics* 15 (May). https://doi.org/10.7238/o.n15.2104.

Roelofsen, Maartje, and Kiley Goyette. 2022. "Second Shift 2.0. Intensifying Housework in Platform Urbanism." In *Platformisation of Urban Life. Towards a Technocapitalist Transformation of European Cities*, edited by Anke Strüver and Sybille Bauriedl. Transcript Verlag.

SABRE. 2017. "The SABRE Story." https://www.sabre.com/files/Sabre-History-rev2017.pdf.

Sadowski, Jathan. 2019. "When Data Is Capital: Datafication, Accumulation, and Extraction." *Big Data and Society* 6 (1): 1–12. https://doi.org/10.1177/2053951718820549.

Schor, Juliet B. 2017. "Does the Sharing Economy Increase Inequality within the Eighty Percent?: Findings from a Qualitative Study of Platform Providers." *Cambridge Journal of Regions, Economy and Society* 10 (2): 263–79. https://doi.org/10.1093/cjres/rsw047.

Srnicek, Nick. 2016. *Platform Capitalism*. Cambridge, UK: Polity.

Timothy, Dallen J, and Victor B Teye. 2009. *Tourism and the Lodging Sector*. Butterworth-Heinemann/Elsevier. https://discovery.uoc.edu/iii/encore/record/C__Rb1039150__STourismandtheLodgingSector__Orightresult__U__X7?lang=eng.

Truitt, Lawrence J., Victor B. Teye, and Martin T. Farris. 1991. "The Role of Computer Reservations Systems. International Implications for the Travel Industry." *Tourism Management* March: 21–36.

Yano, Christine Reiko. 2011. *Airborne Dreams: "Nisei" Stewardesses and Pan American World Airways*. Duke University Press.

CHAPTER 3

The Socio-Spatial Impacts of *Airbnb*

Abstract This chapter provides an overview of the growing impact of short-term rental platforms on urban life, and more generally, their potential to disrupt and deregulate existing order in society and the tourism economy. It argues that the conditions that allow platforms to flourish are contingent upon the different political and sectoral contexts in which they operate and the public interests that are at stake. Moreover, they depend upon governments' interest and ability to enforce (new) regulations upon them. By means of illustration, the chapter describes the development of *Airbnb* in Sofia, Bulgaria and sheds light on the broader urban transformations that have taken place in the city since the end of the socialist regime in 1989. The section provides empirical evidence of the unevenly distributed economic benefits produced by *Airbnb*, further adding to existing socio-economic disparities within the context of Sofia, Bulgaria.

Keywords Platform impacts · Tourism gentrification · Hotel industry · *Airbnb* · Sofia · Bulgaria

© The Author(s), under exclusive license to Springer Nature
Switzerland AG 2022
M. Roelofsen, *Hospitality, Home and Life in the Platform Economies
of Tourism*, https://doi.org/10.1007/978-3-031-04010-8_3

Short-Term Rental Platforms and Their Impacts

This chapter provides an overview of the growing impact of short-term rental platforms on urban life, and more generally, their potential to disrupt and deregulate existing order in society and the tourism and hospitality sector. Among other short-term rental platforms, *Airbnb* has notably received the most scrutiny in academic and popular debates. This is not to say that other short-term rental platforms with similar business models—like Wimdu and VRBO—have not been disruptive or impactful. Yet, *Airbnb* by far dominates the short-term rental market, with a share of 80–95% in many European cities and as such has been held most accountable for the observed socio-economic and spatial impacts (Smigiel 2020; UNWTO 2019). The platform has been associated with a number of pressing issues, but a considerable part of the current critique centres on its contribution to *tourism gentrification*, a process by which certain urban areas, including its residential housing, amenities, and services, have been re-developed and re-purposed to attract, entertain and accommodate short-term visitors and middle-class consumers at the expense of lower income residents (Cocola Gant 2018). In the case of *Airbnb*, these gentrification processes are usually driven by professional actors, rather than hosts who only occasionally rent out the homes in which they live. The platform has in fact become a powerful enabler of new business opportunities for major investors, tourist companies, property managers, and landlords who do not actually live in the *Airbnb*-ed homes that they rent out (Aris Sans and Quaglieri Domínguez 2016; Cocola-Gant and Gago 2021; Gil and Sequera 2020; Semi and Tonetta 2020; Smigiel et al. 2019). Various categories of "professionalism" have been discerned among those renting out properties through the platform (Bosma 2021). Each of these categories contradicts the image that *Airbnb* has persistently promoted of itself as the enabler of the average "home-sharing host" who occasionally rents out a room or home to their guests while being in some way closely involved in their stay as a knowledgeable local (ibid.). Examples of professional hosts are those who operate on a large scale: they usually manage multiple listings on the platform, and (consequently) usually also do not reside in the properties they advertise. These "asset-heavy" hosts are making up an increasingly larger share of hosts on the platform and in some cities, they are responsible for the creation of "ghost hotels", clusters of rooms, or properties that have been bought specifically for short-term rental on platforms like *Airbnb* (Wachsmuth and Weissler 2018). These professional hosts are oftentimes also part of the second category that alludes to professionalism: those who rent out entire properties permanently or for extensive periods of time rather than casually, making hosting a full-time business rather than a source of supplementary income. Another category of professionalism applies to those who outsource the cleaning and hospitality labour to others, or, vice-versa, those who act as intermediaries between property owners and guests, providing "property management services". Processes of professionalization generate greater inequality between those who occasionally rent out the homes or rooms in which they live, and those who engage in *Airbnb* professionally. Bosma (2021, p. 12) argues that asset-heavy hosts "are able to generate a more reliable income stream due their flexibility to switch to other business models and ameliorate risks by reinvesting their capital if needed. This type of privileged professionalization allows property owners to not just generate an income but also to accumulate wealth, for example by benefitting from value appreciation of the apartments they rent out".

For these professionals, the rental income coming from touristic use (based on daily payment) will likely be higher than monthly based payments by long-term residents, therefore, renting out to tourists or other short-term visitors is considered far more lucrative (Yrigoy 2019; Cocola-Gant and Gago 2021). The high turnover of tourists and short-term visitors who stay in these *Airbnb*-ed properties gives these professional actors the opportunity to constantly speculate on rental prices (Cocola-Gant and Gago 2021). In many cities, existing housing and land-use regulations do not permit the conversion of residential housing into short-term rentals, but this has not withheld professional actors from doing so (Wachsmuth and Weisler 2018, p. 1149). Turning residential long-term housing into short-term accommodation has had adverse impacts on both the availability and affordability of urban housing: as housing stock is reallocated from long-term to short-term markets even fewer housing comes available to residents who often already face housing shortages. Secondly, in areas that have undergone tourism gentrification, decreased housing availability and the intensification of land use for touristic purposes have resulted in the rise of property and rental prices (Horn and Merante 2017; Lee 2016). Together, these processes have effectively contributed to the direct and indirect displacement of residents in these areas for whom these areas have become unaffordable (Cocola-Gant 2016; Zanini 2017). In cities that already experience severe housing stress, the *Airbnb* effect has thus been felt like a serious threat to the livelihoods of local population. Various studies have provided compelling empirical evidence of the "*Airbnb* effect" in popular European tourist destinations such as Barcelona, London, Madrid, Paris, Reykjavík, Rome, Venice and Vienna (see for example Arias Sans and Quaglieri Domínguez 2016; Arias Sans et al. 2022; Celata and Romano 2020; Cocola-Gant and Pardo 2018; Ferreri and Sanyal 2018; Freytag and Bauder 2018; Gil and Sequera 2020; Gutiérrez et al. 2017; Mermet 2017; Nofre et al. 2018). Other studies have revealed how the platform affects mid-sized cities in Europe with highly diversified economies that are not necessarily "over-touristified" yet (Ioannides et al. 2019).

Despite this knowledge of *Airbnb*'s negative impact on housing, the platform enterprise has continued to incentivize people to rent out residential homes and rooms to secure profit. It has even gone so far as to provide information to hosts who wish to refinance their mortgages or purchase new properties with "*Airbnb*-friendly" mortgages offered by banks that permit them to be rented out through the platform (Roelofsen 2021). The association of housing and home with financial gains pervades the platform's rhetoric in other ways: it often leverages the advent of the "Great Recession" in 2008 (the same year as *Airbnb* was founded) in its marketing campaigns to argue that it has helped "struggling homeowners" hit by the crisis "to make ends meet". Yet, major factors that invoked the crisis were precisely the financialization *of* and speculation *on* housing (Martin 2011) and it is a similar kind of thinking that underpins *Airbnb*'s marketing campaigns: that, rather than being a human right, housing can be *used* as an object for investment from which profit can be made.

DAILY DISRUPTIONS OF EVERYDAY LIFE

An important characteristic of tourism gentrification is that the process is not necessarily confined to those areas that are initially conceived for tourism, such as city centres or areas around tourist attractions (Cocola-Gant 2018, p. 281). Tourism gentrification can occur in places and neighbourhoods that already provide a "middle class sense of place" with specific consumption facilities. These areas usually also attract tourists, suggesting that processes of tourism and gentrification enforce one another (ibid., p. 282). Neighbourhoods in which *Airbnb* now operates and that were not initially considered or conceived of as major tourist attractions, now provide for what has been termed "new urban tourism".

They appeal to tourists for the same qualities that make these neighbour-hoods attractive as places to live, work, and consume (Novy 2010, p. 31). This (relatively) new form of tourism centres on the experience of "every-day" or "ordinary" city life, relying as such on "the amenities, the retail and entertainment infrastructure that city residents also prefer" (Füller and Michel 2014, p. 1314).

Relatedly, *Airbnb* has contributed to the social displacement of resi-dents through the daily disruptions caused by the increased presence of tourists in residential neighbourhoods and in apartment buildings, or, condominiums (Cocola-Gant 2016; Gurran and Phibbs 2017; Nofre et al. 2018; Zanini 2017). Tourists who stay in residential areas have used public space and accommodation differently than local residents and are sometimes unaware or unobservant of formal and informal agreements on how those spaces are to be collectively used. Complaints among resi-dents pertain to the inappropriate disposal of garbage, parking issues, or ignorance of security, fire, and safety protocols (Gurran and Phibbs 2017). In other instances, inappropriate and hedonistic tourist behaviour have contributed to diminishment of a "sense of place" and belonging among neighbourhood residents. Such behaviour consists of excessive alcohol consumption, and related noise and privacy impacts during the night-time. Other forms of displacement have been provoked by the development of commercially led touristic experiences in public space and privatization, at the expense of already existing (leisure) activities of local populations in those same spaces (Nofre et al. 2018). Neigh-bours who live in close vicinity of *Airbnb*-ed homes have sometimes found themselves unwillingly implicated in the emotional labour of hospitality, by acting as welcoming, "gregarious natives" who offer suggestions for activities in the area (Spangler 2020). In relation to this, neighbours of *Airbnb*-ed homes but also *Airbnb* household members who are not neces-sarily interested in *Airbnb* activity, have had to deploy ad hoc strategies to avoid co-presence or even intimacy with *Airbnb* guests (Roelofsen 2018a; Spangler 2020).

The negative impact of *Airbnb* on local livelihoods has not remained unchallenged, and contestation among local populations has been partic-ularly strong in highly frequented "tourist cities" where the "quality" of the city itself has long been a commodity, consisting of both mate-rial and non-material, symbolic dimensions (Fainstein and Gladstone 1999). Decades of excessive tourism growth—due to increasing demand for mobility, leisure, and unique experiences—have resulted in increased pressure on facilities that are tightly interwoven with other urban (infra)structures and are used by both tourists and residents (Fainstein and Judd 1999). These forms of overcrowding have caused permanent changes to the lifestyles of residents in these places, also termed "over-tourism" (Milano et al. 2019). The touristification of places around the world has led to an increased politicization of tourism "from below"; grass-roots activism which has manifested itself in various ways (Colomb and Novy 2017, p. 20). Residents marching and protesting in the streets (Arias Sans and Quaglieri Domínguez 2016; Cocola-Gant 2016; Colomb and Novy 2017); Twitter hashtag campaigns and other forms of digital activism, which make appeals to local government to push "for stronger regulatory responses to keep *Airbnb*'s practices under control" (Wilson et al. 2022, p. 1095); graffiti on buildings stating that "*Airbnb* [should] die" (Bukowski 2019); and local government campaigns that promise to prohibit *Airbnb* rentals in certain city districts (Ernsting 2020). These initiatives take issue with *Airbnb* and tourism growth more generally and call for a regulation of the tourism economy and changes in urban development and planning models, effectively putting the rights of residents above the perils of capitalism (Colomb and Novy 2017).

AIRBNB'S IMPACT ON THE TOURISM AND HOSPITALITY SECTOR

Another much-discussed topic in both popular and academic literature is *Airbnb*'s impact on the tourism and hospitality sector. Research has demonstrated that platforms like *Airbnb* are used as a substitute for existing accommodations such as budget and mid-range hotels (Guttentag and Smith 2017; Fang et al. 2016; Oskam and Boswijk 2016; Sigala 2017). A resulting quantifiable negative impact on local hotel revenues has been observed in places that have seen a rise in *Airbnb* listings. Hotel lobbies and other established actors in this sector have accused the platform of avoiding zoning laws, evading occupancy taxes, and dismissing public health regulations, all of which are compulsory conditions for hotels and other traditional forms of tourism accommodation to operate (Lee 2016, p. 233). Similarly, *Airbnb* has undermined the protections and regulatory frameworks that pertain to tourism and hospitality labour. Michael O' Regan and Jaeyeon Choe (2017, p. 4) argue that, while *Airbnb* lowers the typical start-up friction costs for hosts, "there are few protections like health coverage, insurance against injuries, paid vacations, pensions, maximum working hours, a stable income, job security, and other safeguards for those hosting via *Airbnb* and those working in the broader *Airbnb* ecosystem".

In relation to this, the COVID-19 (coronavirus) pandemic has provoked changes to the platform's infrastructure and protocols, which allow for easier exposure, extraction, and appropriation of data related to *hospitality work*—and in particular to cleaning and in-person interactions. Recent enforcement of standardized forms of home sanitization has provoked a level of control that is, to some extent, comparable to one employed in commercial hospitality was implemented. Hosts are now encouraged to outsource labour to *Airbnb*-endorsed professional cleaning services companies, which may become a point of controversy for hosts who (can only afford to) clean themselves. This is likely to change the contractual relations and power dynamics between *Airbnb* and both property management companies and professional cleaning companies, which have previously been able to set their own terms and conditions of work in the short-term rental business. These developments also raise important questions about how the standardization and subcontracting of labour through *Airbnb* will trouble existing (national) legislation and guidelines regarding fair work and forms of labour protection in tourism (O' Regan and Choe 2017; Roelofsen and Minca 2021; Ioannides et al. 2021). The scrutinization of *Airbnb* working conditions and compliance with laws and legislation regarding labour is particularly pressing given the existing "prevalence of low pay, exploitation and weak protections across the tourism and hospitality industries" (Bianchi and de Man 2021).

Enforcing Rules and Regulations

Like other platform economies, *Airbnb* is thus a disruptive innovation: it disrupts and deregulates the existing order within society and established sectors (van Dijck et al. 2018). In response to the adverse impacts on local housing markets, local livelihoods, and the hotel industry, some local and national governments have sought to bring short-term-rental platforms like *Airbnb* within the scope of regulation to protect public interests (Frenken et al. 2020). Scholars have argued that "relationships between different government levels and the relative strength of existing policy instruments", are key to understanding evolving regulatory responses to short-term rental platforms such as *Airbnb* (Wilson et al. 2022, p. 1088). In Europe, several types and levels of general regulation have been discerned: firstly, regulation that is applied at the national or state level, secondly, regulation applied at the regional or provincial level, and, thirdly, regulation applied at the local level (mostly municipal or city level). Only recently have there been consultations on the application of pan-European Union regulations to short-term rental platforms (European Commission 2021). These different levels of regulation usually draw on and extend from existing regulations regarding tourism accommodation, housing (zoning), or taxation (Ojamäe et al. 2021; Wilson et al. 2022). Analyses of the different regulatory approaches around the world suggest that they depend on the dominant political-ideological leanings of local governments (Wilson et al. 2022) and that most local governments are relatively lenient towards accommodation platforms (Nieuwland and Melik 2020). Many governments struggle to situate *Airbnb* and other short-term rental platforms into the wider frame of housing or economic policies (Smigiel 2020). With very few exceptions, there are almost no local governments that completely prohibit *Airbnb*'s activities. Instead, popular measures pertain to the limitation of the number of guests in an *Airbnb*-ed property and/or decreeing a minimum or maximum number of nights per year a property can be booked. Other approaches entail limiting or prohibiting *Airbnb* listings in specific neighbourhoods or prohibiting entire property rentals and allowing only for rooms to be booked. Increasingly, governments have required mandatory registration of *Airbnb* listings in municipal registers and/or have successfully imposed tourist occupancy taxes on *Airbnb* rentals (Lee 2016; Arias Sans et al. 2022). In other instances, municipalities require advertised listings to be the host's primary residency, or "owner-occupancy" as a condition (Gurran and Phibbs 2017). Shirley Nieuwland and Rianne van Melik (2020) argue that these "regulations are mostly directed to mitigate neighbourhood impacts, rather than creating a level playing field for the traditional lodging industry". *Airbnb*'s operations pose a set of unique challenges to urban planning and governance and to policymakers (Ferreri and Sanyal 2018; Gurran and Phibbs 2017; Lee 2016). Challenges that are not easily overcome, considering the ongoing evolution of the business models and digital architecture that underpin platforms like *Airbnb*. Sarah Barns argues that these evolutions are by no means accidental or spontaneous but part of a broader set of strategic interventions to exert and maintain control over their markets (2020). Since their inception, many platform companies like *Airbnb* have resisted sharing georeferenced and non-anonymized data with governments, preventing listings to be more effectively monitored, governed, and regulated by authorities (Schor 2017; Ferreri and Sanyal 2018). Only in March 2020 did the European Commission reach an agreement with several platforms, including *Airbnb*, on being permitted to access the platforms' data. However, these data would not identify property owners but only give insight into the number of nights booked and the number of guests, aggregated at the level of municipalities (European Commission 2020).

EXPLORING THE SOCIO-SPATIAL IMPACTS OF *AIRBNB* IN SOFIA, BULGARIA

In the remainder of this chapter, I will discuss Bulgaria's capital Sofia as a case study of *Airbnb*'s operations within Eastern Europe. I provide empirical evidence of the development of *Airbnb* in Sofia's districts, whilst considering some of the broader processes of urban transformation that have taken place in the city in the last decades. Although Sofia receives a moderate number of tourists compared to other European cities that have been studied in relation to the "*Airbnb* effect" (Adamiak 2019), it has allegedly been witnessing an "unprecedented tourism boom" and an increased demand for short-term rental accommodation in the last 5 years (Baltova and Vutsova 2021; UNWTO 2016). The local government and the Ministry of Tourism have generally applauded this boom, as it fits in their broader plans to attract foreign companies, entrepreneurs, and startups to invest in the city (SIA 2018). However, there is also a burgeoning sense that tourism development needs to unfold "sustainably" in the country, and that platform economies like *Airbnb* have a role to play in this (European Council 2018). During an Informal Meeting between the European Ministers of Tourism held in Sofia in February 2018, a European directive was discussed to regulate the flourishing short-term rental platforms, including *Airbnb* (ibid.). Estimates of *Airbnb* listings in Bulgaria remained largely unclear at the time, while the majority of listings on these platforms were not registered with the Ministry of Tourism as official tourism accommodations (Neychev 2017; Dimitrov 2018). This has effectively allowed accommodation providers on the platforms to evade tourist taxes on overnight stays and dismiss the regulations that other actors in the industry have to comply with (ibid.).

However, this case study argues that the rise of *Airbnb* in Sofia needs to be considered beyond its effects on the hospitality industry and its non-compliance with regulations. It wishes in fact to reflect on some of the broader spatial processes of transformation that have occurred in the city since the end of the socialist regime in 1989 and investigates how platforms like *Airbnb* are imbricated in or exacerbate such processes. In doing so, this study relies on an analysis of listing data that were extracted from the platform in 2015 and in 2018. First, I take stock of the diffusion and concentration of *Airbnb* listings in Sofia's neighbourhoods against the backdrop of broader urban transformations in the city's districts. Here I also examine how *Airbnb* contributes to a supposedly more "diversified" offering of tourism accommodation vis-à-vis officially registered tourism accommodation in these same neighbourhoods. Then, I briefly discuss who potentially reaps the benefits and profits from the *Airbnb* economy in Sofia *and* who does not. While mapping out the listings and the hosts' profiles, I simultaneously reflect on some of *Airbnb*'s claims about its alleged positive economic impacts on "communities" worldwide (see Airbnb 2018; but also Roelofsen and Minca 2018).

Sofia

Sofia is the heart of Bulgaria's economic, political and cultural life and the administrative centre of Sofia Province. Demographically it is the most densely populated city in the country, with 1.3 million inhabitants, corresponding to 17.5% of the national population (2011 census data from NSI 2012). During the socialist era, the state took on an important role in reducing housing inequality by means of specific planned urban policies (Vesselinov 2004). Dwellings were allocated to residents through "a socialist administrative method of distribution in accordance with housing need regardless of income" (ibid., p. 2610). But the state also produced what has been termed as an urban *regime of controlled uniformity* through the application of a relatively constant and uniform system of property rights and ensured regulated access to housing loans through the State Savings Bank (ibid., pp. 2610–2611). After the collapse of the socialist regime in 1989, Bulgaria experienced a drastic reduction of the role of the state in all branches of the economy. This change was particularly felt in the housing construction industry and led to a substantial restructuring of the residential market and the patterns of habitation in Sofia (Hirt and Stanilov 2007, p. 218). During the early 1990s, Bulgaria initiated a massive privatization of the housing sector, both in terms of production and ownership. By 2001, 96% of all dwelling units in urban areas were in private hands (ibid.). The transition to a market economy also brought about new sources of housing inequality, such as a "lack of state subsidies for housing, [...] the transfer of responsibility for social housing from the state to the municipalities and changes in mortgage lending" (Vesselinov 2004, p. 2613).

In June 2017 the World Bank published a report on Bulgaria's Housing Sector Assessment that raised a number of serious concerns about Sofia's housing sector (World Bank 2017). The report states that Bulgaria has extremely high vacancy rates and that affordability has become an increasing problem across the country. Household mobility in Bulgaria is one of the lowest among the countries that have transitioned to a market-based economy since the fall of the Iron Curtain (ibid., p. 1). Immediately after the 2007–8 global financial crisis, house prices have dropped by 30–50%, but have now again reached pre-crisis levels, with average market prices being the highest in Sofia City (ibid., p. 107). At the same time, poverty rates have risen. More than a third of young adults continue to live with their parents, as they are unable to afford their own housing. Another major area of concern is vacancy and overcrowding. Over 40% of households live in overcrowded conditions despite a housing vacancy rate of 24% in Sofia. This high vacancy rate might be seen as an opportunity to address the absence of a robust rental market since today less than 5% of the country's housing stock is leased out in the rental market (World Bank 2017, p. 124). However, current regulations in the country tend to erode the rights of both tenants and landlords, proving an important cause for housing market failures (ibid., p. 16; see also Lowe 2003). At the same time, landlords are unwilling to rent their properties out due to the current eviction laws, which, according to the report, tend to favour the tenants. Other aspects that keep landlords from renting out their property are "the inability to enforce formal lease agreements in the court of law" as well as "a general hesitation to rent out the property based on the legacy of poor maintenance" (World Bank 2017, p. 16). Moreover, a "flat 10% income tax applies to rental income, which, although minimal, could be a disincentive to report rental income" (ibid.). For tenants, on the other hand, the problem is the high cost of market-based rentals. Some "42% of single person households, and 31% of tenants in market priced rentals face housing cost over-burdens" (ibid., p. 2). Only the highest income decile (>BGN 1969.00/EUR 1006.00 monthly) of the population can afford high-priced rental properties, whereas less expensive rental properties are only affordable to the seventh income decile and above (i.e. >BGN

1070.00/EUR 547.00 monthly) (ibid.). Up until today, there is little reliable data concerning the rental market in Sofia, its size and dynamics (see also Lowe 2003) making it difficult to say something precise about the actual impact of *Airbnb* in this context. However, recent prospects made by relevant actors in Bulgaria's real estate (financing) continue to be that of a troubled housing and rental housing market for the years to come (Gencheva 2021).

SOFIA'S CITY DISTRICTS AND CLUSTERS

Sofia City is subdivided into 24 districts (see Fig. 3.1), which can be broadly clustered into different "rings" (Hirt and Stanilov 2007). Each of these clusters has gone through their own processes of residential

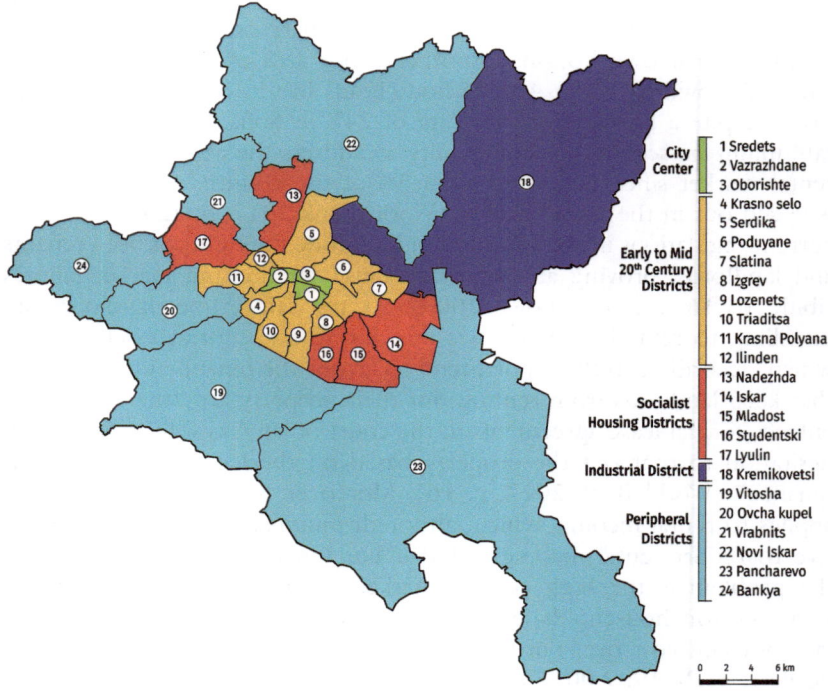

Fig. 3.1 Map of Sofia City districts. Adapted from Hirt (2012, p. 116) and reproduced with the permission of the author (*Source* Originally published in Roelofsen [2018b])

restructuring some of which will be discussed later in conjunction with an analysis of *Airbnb* data.

The first ring is the "historic" central core area with residential buildings built in the late nineteenth and twentieth centuries (Hirt and Stanilov 2007). The central core of Sofia historically hosts the key government, finance, and management functions for the urban region and Bulgaria (Staddon and Mollov 2000, p. 383). It also features a large proportion of the city's official tourist attractions, for example, the Aleksander Nevski Cathedral, the Ancient Serdica Complex, the Regional History Museum, the Archaeological Museum, the Sofia History Museum, the Women's Market, and many more. Since the early 1990s, this ring has gone through a series of spatial transformations, marked by densification, commercialization, and gentrification (Hirt and Stanilov 2007, p. 223). As a consequence of the "unfreezing" of the land and the real estate market after the collapse of the socialist regime, this ring has become an attractive area for foreign investors and exiled Bulgarians to acquire multi-family homes (Brade et al. 2009, p. 228). In the past decade, Sofia's centre has witnessed a conversion of residential spaces to office and commercial use, and the replacement of older structures with larger buildings. This conversion has accommodated primarily commercial and office functions, ostensibly leading to the displacement of lower and middle-income residents (Hirt and Stanilov 2007, p. 223). The city centre has accordingly seen a substantial decline in residents in the decades following the collapse of the socialist regime (Stanilov and Hirt 2014). Moreover, "commercial gentrification", a process by which new luxury shops and services target the more affluent residents, office employees, and visiting tourists, has forced lower rank stores and services to relocate. These lower rank commercial entities provided amenities to lower income residents and the elderly residents of Sofia who, consequently, have been hit hard by the impact of commercial gentrification. This has led to the displacement of vulnerable groups from the inner city to the housing estates located on the city outskirts (ibid., p. 225).

Surrounding the central core are traditional urban neighbourhoods with medium-height residential buildings, which arose during the early and mid-twentieth century (Hirt and Stanilov 2007). Prestigious low-density neighbourhoods in this ring, such as Lozenets, have also undergone intense processes of re-development. They are also the centre of post-socialist gentrification, including high property prices (Hirt 2012, p. 101) and the conversion of public green spaces as well as pre-war single-family housing into upscale (usually gated) medium-height residential housing (Hirt 2006). Here, the first floors along the main streets have been converted to commercial use.

The third ring is the largest and incorporates the vast majority of socialist housing estates, which were built between the early 1960s and the late 1980s. These estates largely consist of modernist high rises made of prefabricated concrete panel blocks, surrounded by shopping and medical facilities, schools, and other services (Hirt and Stanilov 2007; Staddon and Mollov 2000). An estimated 60% of Sofia's residential population lives in such large-scale housing estates (Brade et al. 2009, p. 228). The lack of maintenance of socialist era multi-family buildings represents an area of concern. A reported 75% of these buildings are older than 30 years and suffer from "leaking roofs, damaged facades with fallen plaster, ill-maintained stairwells and hallways, and leaking water and sewer pipes" (World Bank 2017, p. 13).

The fourth ring could be conceived of as an evolving low-density residential periphery, which comprises predominantly single-family homes (Hirt and Stanilov 2007, p. 215). The Vitosha district largely covers the southern part of this ring, which is characterized by a large number of gated communities, constructed by a powerful group of private stakeholders (Brade et al. 2009, p. 228). Smigiel (2013, p. 134) argues that this group was only able to construct these segregated landscapes "because of a neo-liberal policy setting whose main policy pillars are deregulation, decentralisation, privatisation and commodification". Programmes, projects, and strategies of international institutions such as the World Bank and the International Monetary Fund have inspired this urban policy model since the 1990s (ibid.). In the last 20 years, the population in these peripheries has grown exponentially. Vitosha, for example, has witnessed a 60% increase in population and a 150% increase in the number of dwellings (Stanilov and Hirt 2014).

AN EXPLORATION OF *AIRBNB* IN SOFIA

The below-presented analysis is derived from a snapshot of the listings advertised on the platform during May 2015. By using the platform's openly accessible search tool, a general search was conducted using the "Sofia, Bulgaria" search terms, without specifying a specific district or number of persons staying, nor specific periods or a number of nights of stay. The resulting 483 listings that were produced through the search engine were thereafter manually "scraped" after which the data was parsed and categorized. During the first round of analysis, each host was given an ID number in order to differentiate between unique hosts and to account for hosts with multiple listings. The hosts' profile descriptions were thematically analysed using qualitative data analysis software MAXQDA. During the third round of analysis in April 2018, all URL's of the 483 listings that were extracted in 2015 were revisited again to verify: (1) if the May 2015 listings were still active; and (2) how many reviews the listings had received since May 2015.

In 2015, a total of 483 *Airbnb* listings in Sofia showed up in the search results and were analysed. Almost three quarters of all *Airbnb* listings were located in the so-called historical core, and in the southern peripheral districts of the second ring of Sofia (see Fig. 3.2). These are the "culturally desirable districts", which have generally already been subject to processes of commercialization and gentrification since the end of the socialist regime. More specifically, the top 5 highest concentrations of *Airbnb* listings were located within the following districts: Sredets (168 listings); Oborishte (80 listings); Triaditsa (42 listings); Lozenets (38 listings); and Vitosha (22 listings). These together make up for 73% of all *Airbnb* listings in the city. Sredets is the most densely *Airbnb*-listed district and accounts for 35% of all *Airbnb* listings in Sofia. In Sredets 70% of all listings were entire places. In Oborishte, Triaditsa, and Lozenets, over 78% of all listings concerned entire places rather than private or shared rooms within a property. Out of all listings, approximately 73% concerned entire places, 24% were private rooms and the remaining 3% were shared rooms.

Between the initial round of data collection in 2015 and 2018, *Airbnb* listings have increased fivefold in Sofia (SIA 2018). *Sofia Investment Agency*, which is officially part of Sofia Municipality and promotes and assists foreign investment in the city, positively showcases *Airbnb*'s popularity in its most recent report on Tourism and Transport in Sofia (ibid.). Relying on data obtained from *AirDNA*, the report states that *Airbnb* is becoming an increasingly important player in the tourism market, which is reiterated in the report by Boris Pavlov, founder of the local *Airbnb* Property Manager Firm *Flat Manager* (ibid., p. 21). Out of a total of 2353 active *Airbnb* listings in Sofia in 2018, roughly 80% concerned entire places (SIA 2018, p. 20). Returning to the earlier observations made by

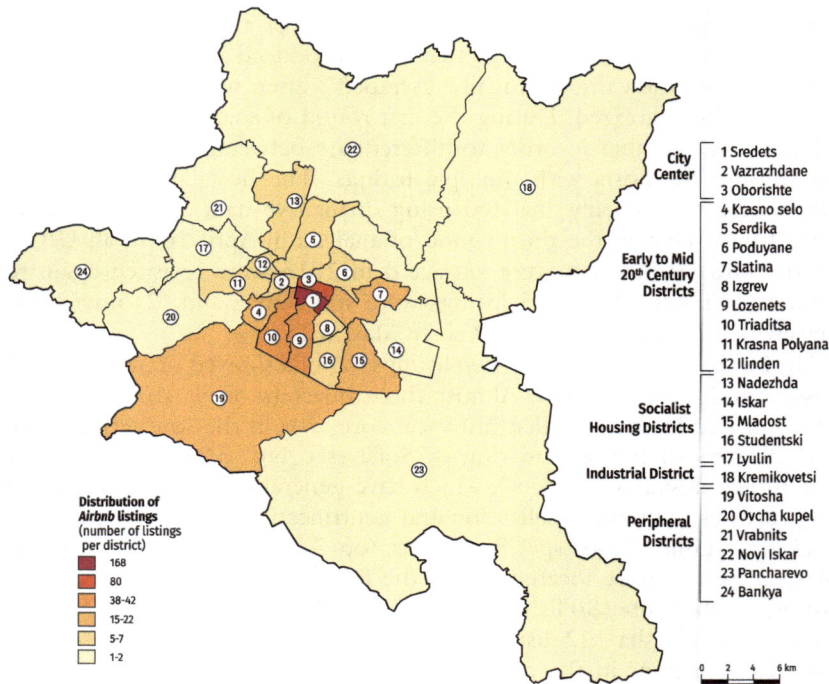

City Center
1 Sredets
2 Vazrazhdane
3 Oborishte

Early to Mid
20th Century
Districts
4 Krasno selo
5 Serdika
6 Poduyane
7 Slatina
8 Izgrev
9 Lozenets
10 Triaditsa
11 Krasna Polyana
12 Ilinden

Socialist
Housing Districts
13 Nadezhda
14 Iskar
15 Mladost
16 Studentski

Industrial District
17 Lyulin
18 Kremikovetsi

Peripheral
Districts
19 Vitosha
20 Ovcha kupel
21 Vrabnits
22 Novi Iskar
23 Pancharevo
24 Bankya

Distribution of Airbnb listings
(number of listings per district)
168
80
38–42
15–22
5–7
1–2

0 2 4 6 km

Fig. 3.2 Distribution of *Airbnb* listings by the district in Sofia (Bulgaria) May 2015 (*Source* The author's own map as originally published in Roelofsen [2018b], compiled from *Airbnb* listing data extracted in May 2015)

the World Bank (2017), the increased presence of *Airbnb* listings in the city, and particularly of entire places rather than shared accommodation, raises questions about what this means for local tenants that currently face an extremely critical rental market. Based on the latest 2011 census data provided by National Statistical Institute (NSI 2012), Sofia had approximately 600,000 dwellings. Although there are no exact figures available on the total dwellings for rent in Sofia, according to the World Bank report (2017) less than 5% of all dwellings—approximately 30,000—were available on the rental market. Considering that those 1882 *Airbnb* listings in Sofia concern entire apartments, one can presume that these make up at least 6% of the current estimated rental market. As it stands,

Airbnb provides local property managers and landlords with the opportunity to rent out to tourists rather than to residents, affording them to avoid dealing with the problematic regulations around rental housing. At the same time, local residents who seek long-term rental housing in an already tight housing market, find themselves confronted with the encouragement of private investment in tourism accommodation rather than long-term secure housing.

When comparing the approximately 1,100 *Airbnb* beds listed in Sofia in 2015 to an approximate 18,400 beds within Sofia's official Accommodation Establishments (also termed "AE's" by Sofia Tourism Administration 2017) a few observations can be made. Just over 55% of all *Airbnb* beds are located within Sredets and Oborishte. These districts are part of the city ring and harbour some of the most commonly visited tourist attractions. While Oborishte has a moderate share of AE-beds, Sredets accounts for a share of over 8% of total AE-beds in Sofia (Sofia Tourism Administration 2016). Another 15% of all *Airbnb* beds are located within Triaditsa and Lozenets, which are both located in the second ring of Sofia. These two districts make up for over 20% of the total AE-beds in Sofia, with Lozenets corresponding to a little over 12% of all beds (ibid.). The Vitosha district, which is located at the southern rims of Sofia and is famous for the Vitosha National Park and Vitosha Mountain, accommodates the largest number of official Accommodation Establishments and the second largest number of AE-beds. While being the fifth most *Airbnb*-listed district, it only accounts for approximately 40 *Airbnb* beds compared to approximately 1800 AE-beds. Although the most densely *Airbnb*-listed districts indeed resemble the most popular AE-districts in Sofia, all official AE's enjoy a slightly more even spread throughout all districts in Sofia. Thus, despite *Airbnb*'s claims that 74% of *Airbnb* properties are located outside the main hotel districts (Airbnb 2018) the concentration of listings and beds in Sofia's city centre and second ring show a rather different picture (see Figs. 3.2 and 3.3).

The somewhat less popular peripheral districts and the socialist housing districts Kremikovtsi, Novi Iskar, Nadezhda, Lyulin and Iskar have the lowest concentration of both *Airbnb* listings and AE's. Krasna Polyana and Ilinden are perhaps the only two districts that do not have AE's and do have 5 and, respectively, 7 *Airbnb* listings each. The densely populated lower- and middle-income districts—roughly clustered within the third ring—are minimally represented in the listing data. Districts such as Mladost, Studentski, Iskar, Lyulin, and Nadezhda which can be clustered

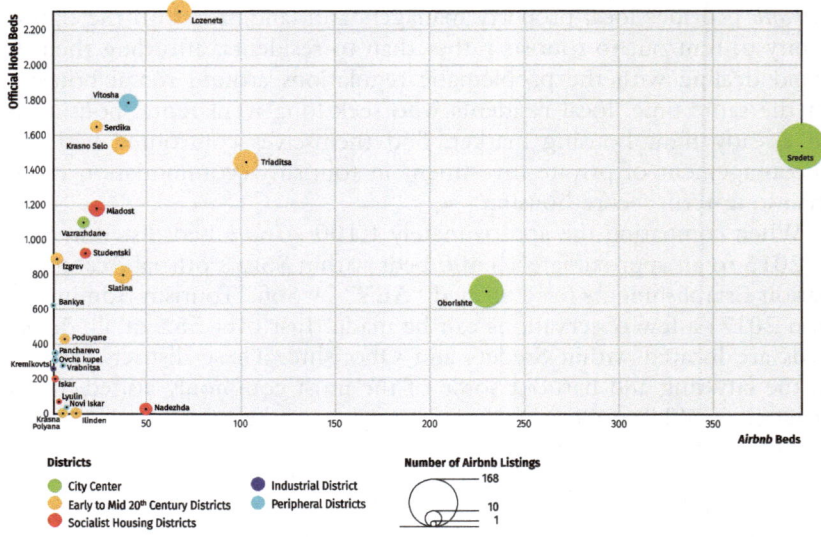

Fig. 3.3 Official hotel beds versus *Airbnb* beds by district in Sofia (Bulgaria) May 2015 (*Source* Originally published in Roelofsen [2018b], compiled from *Airbnb* listing data extracted in May 2015 and data from Sofia Tourism Administration 2016)

under the third ring, have the highest population number and are the largest in terms of area, but represent a mere 7% of all *Airbnb* listings. These districts incorporate the vast majority of socialist housing estates. And although these lower- and middle-income districts could arguably profit from an extra *Airbnb* income, they also represent the least reviewed *Airbnb*-listed districts.

Guest Reviews of *Airbnb* listings in Sofia

A significant limitation in estimating *Airbnb*'s potential economic impact in specific districts comes from the fact that the platform does not disclose information on the number of bookings made by its guests. Although reviews do not elicit information on the number of nights that were booked nor the overnight fees that were charged for the stay, they may however provide some approximate esteem of the popularity of certain districts. Consequently, the reviews may disclose some information on who profits from *Airbnb*, and who does not. Out of all 483 listings in Sofia, 206 listings received reviews, whereas 277 listings did not receive any reviews. A total of 2629 reviews were counted in May 2015. Almost 80% of all reviews could be attributed to listings within three districts: Sredets (1509 reviews, or, 58%), Oborishte (399 reviews, or, 15%), and Lozenets (150 reviews, or, 6%). Again, these are the districts in which we find the main hotel providers, tourist attractions and are the districts with high property and rental prices, experiencing densification, commercialization, and gentrification. This shows how tourism and gentrification in Sofia are effectively interrelated—those areas that already attract middle-class residents and have undergone gentrification processes are most popular among *Airbnb* guests. However, listings in the following districts, largely characterized by social housing estates, did not show any reviews: Poduyane, Iskar, Ovcha kupel, Bankya, Kremikovtsi, Lyulin, Novi Iskar, and Vrabnitsa.

In Sredets, the top 23 listings (out of 87 reviewed listings) received 1016 reviews. These 23 listings thus account for 68% of all reviews in the district. In Oborishte the top 10 listings (out of 37 reviewed listings) received 272 reviews, which account for 68% of the total reviews in the district. And in Lozenets, the top 5 listings (out of 19 reviewed listings) received 113 reviews, which account for 75% of the total reviews in the district. What these top three *Airbnb* districts therefore share is that a little over 25% of reviewed listings make up for almost 70% or more of the total reviews. This makes the distribution of *Airbnb* revenues strongly unequal not just between districts but also within districts in Sofia. Moreover, more than 75% of all reviews account for stays in entire places rather than (shared) bedrooms within people's homes.

Finally, all 2015 *Airbnb* listings in Sofia were revisited again online in April 2018 using the listings' unique URL's. This second round of analysis aimed to verify: (1) if the 2015 listings were still active; and (2) how many reviews these listings had received since 2015. Out of 418 total properties visible on the platform in 2015, 193 properties (46%) were no longer listed in 2018. Out of the remaining properties still listed, 124 properties (55%) still did not receive a single guest review to date. The remaining properties received a total of 5984 extra guest reviews on top of their existing reviews received in 2015. Conforming to the findings from 2015, Sredets, Oborishte, and Lozenets remain the top three districts in terms of guest reviews over a period of 3 years (May 2015–April 2018).

On average, the remaining 2015 listings in Sredets enjoyed a growth of 233%; those in Oberishte of 271%, and those in Lozenets of 267%.

AIRBNB SOFIA'S "LOCAL" HOSTS

A total of 318 unique hosts were counted on the platform. The large majority identified themselves as residents from Sofia, but others as Bulgarian nationals living abroad and a few others as foreign residents. Most profiles/biographies were written in English, several were written in German, Spanish, or French, and only three in Cyrillic. The large majority of hosts declared to be fervent travellers themselves, open to other cultures and were keen to share their "version" of Sofia with *Airbnb* guests. Most profiles also attested to highly educated people who worked as lawyers, IT—and software specialists, artists, designers, journalists, marketing—and PR specialists. This confirms what other studies have found: that *Airbnb* is a relatively exclusive marketplace for highly educated professionals (e.g. Frenken and Schor 2017; Schor 2017) with privileges such as the political right to mobility and the economic means to travel. At first sight, a total of 27% *Airbnb* listings of private and shared rooms in Sofia could be interpreted as rooms that are "shared" within a private property of a local host. However, a more detailed analysis showed a different picture. Out of these 128 private/shared rooms, 35 rooms (27%) were advertised by property managers, hostel owners, hostels, B&B's, and guesthouses.

A total of 250 hosts advertised one listing on the platform and almost half of all *Airbnb* listings in Sofia were advertised by only 68 hosts. Very optimistically assuming that all hosts advertised at least one of the properties in which they usually lived, around one-third of Sofia's *Airbnb* listings represented properties that were either second homes or properties that were not owner-occupied. Furthermore, a total of 32 hosts on the platform were representatives or personnel of commercial entities such as hostels, hotels, B&B's, guesthouses, and rental agencies. For example, host Daniela (a pseudonym) managed 13 entire *Airbnb* listings in Sofia. Daniela's biography explained: "I manage professionally designed apartments in Sofia's top neighbourhoods catered towards discerning travellers (leisure or business) who want to experience Sofia". Professional hosts like Daniela accounted for 92 *Airbnb* listings, or, 19% of the total listings in Sofia. These listings received a total of 247 reviews (10% of the total reviews). It has been argued that some of these commercial entities profit from *Airbnb*'s relatively low booking commissions and the option to rate users in order to encourage appropriate behaviour (Epstein 2014; ITB 2014). Here, *Airbnb* arguably also allows property managers and landlords to ask for higher rents from tourists, whilst potentially not declaring income tax, which is mandatory when renting out to resident tenants. Reflecting on the abovementioned results, one can assume that a significant number of hosts on the platform do not represent the celebrated "local" hosts who "share the home in which they live", as claimed by *Airbnb* in its own reports (2018).

LIMITATIONS

While discussing these findings, however, a number of limitations in this analysis have to be taken into account. The first and foremost limitation pertains to the inherent fluidity and partiality of the data. Since hosts can adjust, list, and unlist their property at their convenience and at any given time, the dataset in this study only captures *Airbnb* listings within a specific time frame. *Airbnb* listings—including entire places—can be advertised on the platform for only a very limited period of time, while being inhabited for the remainder of the time. The findings presented above should therefore be read with caution and at no time should they be referred to without mentioning the fluid and partial nature of the data. The second limitation is that the listings on the platform were not accurately geo-referenced. *Airbnb* did not disclose the exact location of the listings and listings can be anywhere between 0–150 metres away from the actual location on the map that is paired up with *Airbnb*'s search engine. While a few *Airbnb* hosts did disclose the exact street number of their property in their advertisement, the majority of hosts did not. Consequently, the street name of each listing and its location on a map (using the zoom function) was attributed to a specific district in Sofia City as accurately as possible but a margin of 0–150 metres had to be taken into account. The 84 anonymized listings that did not feature street names but instead mentioned "Sofia City" as a location, were each attributed to a district based on their location on the map that is paired up with *Airbnb*'s search engine.

CONCLUSION

This case study aimed to explore the spatialities of *Airbnb* in Sofia, taking into account the broader processes of urban transformation that have developed in the city over the past decades. Despite Sofia attracting a moderate yet rapidly growing number of international tourists compared to other more popular tourist cities in Europe, this study has shown that *Airbnb* had an impact somewhat similar to those experienced in other destinations. However, *Airbnb*'s presence in Sofia also revealed something specific, due to its recent history as a post-socialist city. *Airbnb* listings predominantly concentrate in those districts that have undergone processes of (commercial) gentrification and also harbour the majority of officially registered tourism accommodations and tourism attractions. These are, at the same time, the most affluent areas of Sofia, areas that house the growing middle- and upper-middle classes of Bulgaria (Hirt and Stanilov 2007). Thus, while *Airbnb* declares to offer a "local" experience to travellers by staying in non-tourist-saturated neighbourhoods, the findings in this study suggest otherwise. Commercialization, gentrification, and a privileged higher income population mark the most popular *Airbnb* neighbourhoods in Sofia. In a similar vein, *Airbnb* mainly benefits those hosts who already profit from a privileged position: both in terms of their residential location in the city, as well as in terms of their socio-economic status. Perhaps equally important is the fact that this study has brought to light the socio-spatial inequalities produced by *Airbnb* in relation to larger economic and political processes that have prevailed in the city after the end of the socialist regime. These include the unregulated privatization of the housing market and a neo-liberal housing agenda that resulted in "a delegitimation of all kind of public involvement in urban development", a philosophy still guiding Sofia's urban policy today (Smigiel 2013, p. 128). A philosophy that, ironically enough, was promoted in the past by the same World Bank that now expresses its concerns about the current state of Bulgaria's housing market (ibid.).

The municipality's burgeoning desire for tourism growth and its celebration of the tourist "boom" to incentivize more foreign investment seem to offer the ideal conditions for *Airbnb* business to flourish in Sofia. Although housing security (and affordability) might be pressing concerns for the municipality and Bulgaria more generally, the prioritization of tourism development is not perceived as a process potentially aggravating those same problems. This is somewhat surprising as Bulgaria is certainly not new to the unique set of challenges that the "sharing economy" brings to urban governance and policymakers. A far more radical stance was taken when car-service platform *Uber* first launched in Bulgaria in 2014 (Gavrilov 2015). The platform was quickly met with massive opposition from taxi drivers and state institutions, who claimed that "neither the company [*Uber*] nor the drivers comply with the road transport law and other relevant legislation" (Markova 2016). About one year later, Bulgaria's Supreme Court forced the platform to seize its operations. But while *Airbnb* has operated in Bulgaria for much longer than *Uber*, the aforementioned Informal Meeting between the EU Ministers of Tourism on curbing *Airbnb* within the country is a relatively new one (Dimitrov 2018). In order to prevent a further decrease in the city's already critical rental housing stock and the associated spatial transformations, urban planners and policymakers in Sofia may learn from other cities around the world that have regulated the platform (see, for example, Nieuwland and van Melik 2020). At the same time, this work on Sofia will hopefully inspire other studies on cities in Eastern Europe with lower numbers of tourist arrivals where *Airbnb* may have a similar impact on social change in the city.

POSTSCRIPT

Since the completion of this case study on *Airbnb* in Sofia, the socio-political dynamics and regulatory approaches to *Airbnb* in Bulgaria have altered, detailed most recently in a report by Stela Baltova and Albena Vutsova (2021). They contend that, according to the country's Tourism Act, apartments and guest houses are class "B" touristic dwellings and, as such, *Airbnb* accommodations are subject to categorization (Baltova and Vutsova 2021, p. 80). Those who rent out these apartments and guesthouses do not have to be registered under Commercial Law and can operate without registering as a company. In November 2019, senior politicians in the ruling majority came to an understanding that platform economy services such as those provided by *Airbnb* run in breach of the Tourism Act. Accordingly, new amendments to the Act were adopted by the parliament. The amendments now require that those who rent out apartments and rooms through platforms such as *Booking.com*, *Airbnb*, *Expedia*, and *Facebook* groups will only be allowed to do so if registered under the Tourism Act. Moreover, as of the 1st of January 2020, owners who offer accommodation through platforms such as *Airbnb*, will have to pay a patent tax. According to Baltova and Vutsova (2021), the aim of the amended Act is to create conditions and prerequisites for those participating in the platform economy, ensuring they comply with the legislation in that specific economic sector (the tourism sector in this case). This should ensure equal treatment of different providers of tourist services and allows for economic activity to be traced.

REFERENCES

Adamiak, Czesław. 2019. "Current State and Development of Airbnb Accommodation Offer in 167 Countries." *Current Issues in Tourism* (December): 1–19. https://doi.org/10.1080/13683500.2019.1696758.

Airbnb. 2018. "Airbnb's Positive Economic Impact in Cities Around the World." https://blog.atairbnb.com/economic-impact-airbnb/act.

Baltova, Stela, and Albena Vutsova. 2021. "Setting the Stage of the Sharing Economy: The Case of Bulgaria." In *The Collaborative Economy in Action: European Perspectives*, edited by Andrzej Klimczuk, Vida Česnuitytė, and Gabriela Avram, 75–89. University of Limerick, Ireland.

Bianchi, Raoul V., and Frans de Man. 2021. "Tourism, Inclusive Growth and Decent Work: A Political Economy Critique." *Journal of Sustainable Tourism* 29 (2–3): 353–71. https://doi.org/10.1080/09669582.2020.1730862.

Bosma, Jelke R. 2021. "Platformed Professionalization: Labor, Assets, and Earning a Livelihood through Airbnb." *Environment and Planning A: Economy and Space* (January): 0308518X2110634. https://doi.org/10.1177/0308518X211063492.

Brade, Isolde, Günter Herfert, and Karin Wiest. 2009. "Recent Trends and Future Prospects of Socio-Spatial Differentiation in Urban Regions of Central and Eastern Europe: A Lull Before the Storm?" *Cities* 26 (5): 233–44. https://doi.org/10.1016/j.cities.2009.05.001.

Bukowski, Wolf. 2019. "Viva Airbnb, Crepino Gli Inquilini!" Jacobin Italia. 2019. https://jacobinitalia.it/viva-airbnb-crepino-gli-inquilini/.

Celata, Filippo, and Antonello Romano. 2020. "Overtourism and Online Short-Term Rental Platforms in Italian Cities." *Journal of Sustainable Tourism* 30 (5): 1020–39. https://doi.org/10.1080/09669582.2020.1788568.

Cocola-Gant, Agustín. 2016. "Holiday Rentals: The New Gentrification Battlefront." *Sociological Research Online* 21 (3): 112–20. https://doi.org/10.5153/sro.4071.

Cocola-Gant, Agustín. 2018. "Tourism Gentrification." In *Handbook of Gentrification Studies*, 281–93. Edward Elgar Publishing. https://doi.org/10.4337/9781785361746.00028.

Cocola-Gant, Agustin, and Ana Gago. 2021. "Airbnb, Buy-to-Let Investment. and Tourism-Driven Displacement: A Case Study in Lisbon." *Environment and Planning A: Economy and Space* 53 (7): 1671–88. https://doi.org/10. 1177/0308518X19869012

Cocola-Gant, Agustín, and Daniel Pardo. 2018. "Resisting Tourism Gentrifica- tion. The Experience of Grassroots Movements in Barcelona." *Urbanistica Tre, Giornale Online di Urbanistica* 5 (13): 39–47.

Colomb, Claire, and Johannes Novy. 2017. *Protest and Resistance in the Tourist City*. London and New York: Routledge.

Dijck, José Van, Thomas Poell, and Martijn De Waal. 2018. *The Platform Society: Public Values in a Connective World*. Oxford University Press.

Dimitrov, Martin. 2018. "Bulgaria Minister Takes Aim at Airbnb, Booking.Com." Balkan Insight. http://www.balkaninsight.com/en/art icle/bulgarian-tourism-ministers-wants-airbnb-booking-regulated-by-eu-02- 22-2018.

Epstein, Eli. 2014. "Hostels Embrace Airbnb in Effort to Escape Rising Booking Fees." Mashable. http://mashable.com/2014/07/19/airbnb-hos tels/Iw_z5y3MVGqY.

Ernsting, Zeeger. 2020. "De Binnenstad Moet Weer Een Levende Stadswijk Worden." Groenlinks. https://amsterdam.groenlinks.nl/nieuws/de-binnen stad-moet-weer-een-levende-stadswijk-worden.

European Commission. 2020. "Airbnb, Booking, Expedia and Tripadvisor to Share Data with Eurostat." https://ec.europa.eu/commission/presscorner/ detail/en/ip_20_194.

European Commission. 2021. "Tourist Services—Short-Term Rental Initia- tive." https://ec.europa.eu/info/law/better-regulation/have-your-say/initia tives/13108-Tourist-services-short-term-rental-initiative_en.

European Council. 2018. "Arrival and Doorstep Bulgarian EU Presi- dency (Angelkova) Video." https://newsroom.consilium.europa.eu/permal ink/194738.

Fainstein, Susan S., and David Gladstone. 1999. *Evaluating Urban Tourism. The Tourist City*. New Haven, CT: Yale University Press.

Fang, Bin, Qiang Ye, and Rob Law. 2016. "Effect of Sharing Economy on Tourism Industry Employment." *Annals of Tourism Research* 57 (January 2013): 264–67. https://doi.org/10.1016/j.annals.2015.11.018.

Ferreri, Mara, and Romola Sanyal. 2018. "Platform Economies and Urban Plan- ning: Airbnb and Regulated Deregulation in London." *Urban Studies* 55 (15): 3353–68. https://doi.org/10.1177/0042098017751982.

Frenken, Koen, and Schor Juliet. 2017. "Putting the Sharing Economy into Perspective." *Environmental Innovation and Societal Transitions*: 233–10. https://doi.org/10.1016/j.eist.2017.01.003

Frenken, Koen, Arnoud van Waes, Peter Pelzer, Magda Smink, and Rinie van Est. 2020. "Safeguarding Public Interests in the Platform Economy." *Policy & Internet* 12 (3): 400–425. https://doi.org/10.1002/poi3.217.

Freytag, Tim, and Michael Bauder. 2018. "Bottom-up Touristification and Urban Transformations in Paris." *Tourism Geographies* 20 (3): 443–60. https://doi.org/10.1080/14616688.2018.1454504.

Füller, Henning, and Boris Michel. 2014. "'Stop Being a Tourist!' New Dynamics of Urban Tourism in Berlin-Kreuzberg." *International Journal of Urban and Regional Research* 38 (4): 1304–18. https://doi.org/10.1111/1468-2427.12124.

Gavrilov, Peter. 2015. "After UBER, Is Airbnb OK?" Webcafe. http://www.webcafe.bg/webcafe/parite/id_1873048908.

Gencheva, Boryana. 2021. "Пет Прогнози За Пазара На Жилища През 2022 Г." *Kapital* (2021). https://www.capital.bg/biznes/imoti/2021/12/23/4294142_pet_prognozi_za_pazara_na_jilishta_prez_2022_g/.

Gil, Javier, and Jorge Sequera. 2020. "The Professionalization of Airbnb in Madrid: Far from a Collaborative Economy." *Current Issues in Tourism*: 1–20. https://doi.org/10.1080/13683500.2020.1757628.

Gurran, Nicole, and Peter Phibbs. 2017. "When Tourists Move In: How Should Urban Planners Respond to Airbnb?" *Journal of the American Planning Association* 83 (1): 80–92. https://doi.org/10.1080/01944363.2016.1249011.

Gutiérrez, Javier, Juan Carlos García-Palomares, Gustavo Romanillos, and María Henar Salas-Olmedo. 2017. "The Eruption of Airbnb in Tourist Cities: Comparing Spatial Patterns of Hotels and Peer-to-Peer Accommodation in Barcelona." *Tourism Management* 62: 278–91. https://doi.org/10.1016/j.tourman.2017.05.003.

Guttentag, Daniel A., and Stephen L. J. Smith. 2017. "Assessing Airbnb as a Disruptive Innovation Relative to Hotels: Substitution and Comparative Performance Expectations." *International Journal of Hospitality Management* 64: 1–10. https://doi.org/10.1016/j.ijhm.2017.02.003.

Hirt, Sonia. 2006. "Post-Socialist Urban Forms: Notes From Sofia." *Urban Geography* 27 (5): 464–88. https://doi.org/10.2747/0272-3638.27.5.464.

Hirt, Sonia. 2012. *Iron Curtains: Gates, Suburbs and Privatization of Space in the Post-Socialist City*. Malden: Wiley.

Hirt, Sonia, and Kiril Stanilov. 2007. "The Perils of Post-Socialist Transformation: Residential Development in Sofia BT." In *The Post-Socialist City: Urban Form and Space Transformations in Central and Eastern Europe After Socialism*, edited by Kiril Stanilov, 215–44. Dordrecht: Springer Netherlands. https://doi.org/10.1007/978-1-4020-6053-3_11.

Horn, Keren, and Mark Merante. 2017. "Is Home Sharing Driving up Rents? Evidence from Airbnb in Boston." *Journal of Housing Economics* 38: 14–24. https://doi.org/10.1016/j.jhe.2017.08.002.

Ince, Anthony, and Sarah Marie Hall. 2018. *Sharing Economies in Times of Crisis: Practices, Politics and Possibilities*. Routledge.

Ioannides, Dimitri, Michael Röslmaier, and Egbert van der Zee. 2019. "Airbnb as an Instigator of 'Tourism Bubble' Expansion in Utrecht's Lombok Neighbourhood." *Tourism Geographies* 21 (5): 822–40. https://doi.org/10.1080/14616688.2018.1454505.

Ioannides, Dimitri, Szilvia Gyimóthy, and Laura James. 2021. "From Liminal Labor to Decentwork: A Human-Centered Perspective on Sustainable Tourism Employment." *Sustainability (Switzerland)* 13 (2): 1–15. https://doi.org/10.3390/su13020851.

ITB. n.d. "ITB World Travel Trends Report 2014/2015."

Lee, Dayne. 2016. "How Airbnb Short-Term Rentals Exacerbate Los Angeles's Affordable Housing Crisis: Analysis and Policy Recommendations Student Notes." *Harvard Law & Policy Review* 10 (1): 229–54. https://heinonline.org/HOL/P?h=hein.journals/harlpolrv10&i=235.

Lowe, Stuart. 2003. "The Private Rented Sector—Evidence from Budapest and Sofia." In *Housing Change in East and Central Europe: Integration or Fragmentation?*, edited by Stuart Lowe and Sasha Tsenkova, 63–72. London: Routledge.

Markova, Ekaterina. 2016. "Bulgaria: Supreme Court Shuts down Smartphone Car Service Uber." *Eurofound* (2016). https://www.eurofound.europa.eu/publications/article/2016/bulgaria-supreme-court-shuts-down-smartphone-car-service-uber.

Martin, Ron. 2011. "The Local Geographies of the Financial Crisis: From the Housing Bubble to Economic Recession and Beyond." *Journal of Economic Geography* 11 (4): 587–618. http://www.jstor.org/stable/26162231.

Mermet, Anne-Cécile. 2017. "Critical Insights from the Exploratory Analysis of the 'Airbnb Syndrome' in Reykjavík." In *Tourism and Gentrification in Contemporary Metropolises: International Perspectives*, edited by Maria Gravari-Barbas and Sandra Guinand, 52–74. New York: Routledge.

Milano, Claudio, Marina Novelli, and Joseph M Cheer. 2019. "Overtourism and Tourismphobia: A Journey Through Four Decades of Tourism Development, Planning and Local Concerns." *Tourism Planning & Development* 16 (4): 353–57. https://doi.org/10.1080/21568316.2019.1599604.

Neychev, Nikolai. 2017. "Airbnb or Long Term Rental." *Kapital* (April 2017). https://www.capital.bg/biznes/moiat_kapital/2017/04/08/2949437_a irbnb_ili_dulgosrochen_naem/.

Nieuwland, Shirley, and Rianne van Melik. 2020. "Regulating Airbnb: How Cities Deal with Perceived Negative Externalities of Short-Term Rentals."

Current Issues in Tourism 23 (7): 811–25. https://doi.org/10.1080/136 83500.2018.1504899.

Nilsson, Jan Henrik, Stefan Gössling, Szilvia Gyimóthy, and Carlo Aall. 2019. "Klickturismens Baksida Borde Bekymra Oss." *Svenska Dagbladet, Stockholm,* 4.

Nofre, Jordi, Emanuele Giordano, Adam Eldridge, João C Martins, and Jorge Sequera. 2018. "Tourism, Nightlife and Planning: Challenges and Opportunities for Community Liveability in La Barceloneta." *Tourism Geographies* 20 (3): 377–96. https://doi.org/10.1080/14616688.2017.1375972.

Novy, Johannes. 2010. "What's New About New Urban Tourism? And What Do Recent Changes in Travel Imply for the 'Tourist City' Berlin?" In *The Tourist City Berlin: Tourism and Architecture,* 26–35. Salenstein: Braun. https://opus4.kobv.de/opus4-UBICO/frontdoor/index/index/docId/11047.

NSI. 2012. "Population and Housing Census in the Republic of Bulgaria 2011."

Ojamäe, Liis, Kristjan Peik, Kairit Kall, Triin Roosalu, and Marge Unt. 2021. "Hosting in Airbnb: Platform Work at the Intersection of Hospitality, Accommodation and Home-Making."

O'Regan, Michael, and Jaeyeon Choe. 2017. "Airbnb and Cultural Capitalism: Enclosure and Control Within the Sharing Economy." *Anatolia* 28 (2): 163–72. https://doi.org/10.1080/13032917.2017.1283634.

Oskam, Jeroen, and Albert Boswijk. 2016. "Airbnb: The Future of Networked Hospitality Businesses." *Journal of Tourism Futures* 2 (1): 22–42. https://doi.org/10.1108/JTF-11-2015-0048.

Roelofsen, Maartje. 2018a. "Performing 'Home' in the Sharing Economies of Tourism: The Airbnb Experience in Sofia, Bulgaria." *Fennia* 196 (1): 24–42. https://doi.org/10.11143/fennia.66259.

Roelofsen, Maartje. 2018b. "Exploring the Socio-Spatial Inequalities of Airbnb in Sofia, Bulgaria." *Erdkunde* 72 (4): 313–27. https://doi.org/10.3112/erd kunde.2018.04.04.

Roelofsen, Maartje. 2021. "Capitalizing on Crises. Transformations in Airbnb from the Great Recession through the Covid-19 Pandemic." In *From Overtourism to Undertourism: Sustainable Scenarios in Post Pandemic Times,* edited by Valentina Pecorelli, 33–54. Milan: UNICOPLI.

Roelofsen, Maartje, and Claudio Minca. 2018. "The Superhost. Biopolitics, Home and Community in the Airbnb Dream-World of Global Hospitality." *Geoforum* 91: 170–81. https://doi.org/10.1016/j.geoforum.2018.02.021.

Roelofsen, Maartje, and Claudio Minca. 2021. "Sanitised Homes and Healthy Bodies: Reflections on Airbnb's Response to the Pandemic." *Oikonomics* 15 (May). https://doi.org/10.7238/o.n15.2104.

Sans, Albert Arias, and Alan Quaglieri Domínguez. 2016. "Unravelling Airbnb. Urban Perspectives." In *Reinventing the Local in Tourism,* edited by Paolo Russo and Greg Richards, 209–228. Bristol, UK: Channel View.

Sans, Albert Arias, Alan Quaglieri-Domínguez, and Antonio Paolo Russo. 2022. "Home-Sharing as Transnational Moorings." *City* (January): 1–19. https://doi.org/10.1080/13604813.2021.2018859.

Schor, Juliet B. 2017. "Does the Sharing Economy Increase Inequality within the Eighty Percent?: Findings from a Qualitative Study of Platform Providers." *Cambridge Journal of Regions, Economy and Society* 10 (2): 263–79. https://doi.org/10.1093/cjres/rsw047.

Semi, Giovanni, and Marta Tonetta. 2020. "Marginal Hosts: Short-Term Rental Suppliers in Turin, Italy." *Environment and Planning A: Economy and Space* 53 (7): 1630–51. https://doi.org/10.1177/0308518X20912435.

SIA. 2018. "Sofia Investment Agency: Sofia Tourism and Air Transport Market Report. September 2018." http://investsofia.com/en/introducing-the-new-sofia-tourism-and-air-transport-market-report-2018.

Sigala, Marianna. 2017. "Collaborative Commerce in Tourism: Implications for Research and Industry." *Current Issues in Tourism* 20 (4): 346–55. https://doi.org/10.1080/13683500.2014.982522.

Smigiel, Christian. 2013. "The Production of Segregated Urban Landscapes: A Critical Analysis of Gated Communities in Sofia." *Cities* 35 (December): 125–35. https://doi.org/10.1016/j.cities.2013.06.008.

Smigiel, Christian. 2020. "Why Did It Not Work? Reflections on Regulating Airbnb and the Complexity and Agency of Platform Capitalism." *Geographica Helvetica* 75 (3): 253–57.

Smigiel, Christian, Angela Hof, Karolin Kautzschmann, and Roman Seidl. 2019. "No Sharing! Ein Mixed-Methods-Ansatz Zur Analyse von Kurzzeitvermietungen Und Ihren Sozialräumlichen Auswirkungen Am Beispiel Der Stadt Salzburg." *Raumforschung Und Raumordnung / Spatial Research and Planning* 78 (2): 153–70. https://doi.org/10.2478/rara-2019-0054.

Spangler, Ian. 2020. "Hidden Value in the Platform's Platform: Airbnb, Displacement, and the Un-Homing Spatialities of Emotional Labour." *Transactions of the Institute of British Geographers* 45 (3): 575–88. https://doi.org/10.1111/tran.12367.

Staddon, Caedmon, and Bellin Mollov. 2000. "City Profile." *Cities* 17 (5): 379–87. https://doi.org/10.1016/S0264-2751(00)00037-8.

Stanilov, Kiril, and Sonia Hirt. 2014. "Sprawling Sofia." In *Confronting Suburbanization*, 163–91. Chichester, UK: Wiley. https://doi.org/10.1002/9781118295861.ch6.

UNWTO. 2016. "International Tourism Trends in EU-28 Member States. Current Situation and Forecasts for 2020-2025-2030." https://ec.europa.eu/growth/content/international-tourism-trends-eu-28-member-states-current-situation-and-forecast-2020-2025-0_en.

UNWTO. 2019. "New Business Models in the Accommodation Industry – Benchmarking of Rules and Regulations in the Short-Term Rental

Market, Executive Summary." Madrid. https://www.e-unwto.org/doi/pdf/
10.18111/9789284421190.

Vesselinov, Elena. 2004. "The Continuing 'Wind of Change' in the Balkans:
Sources of Housing Inequality in Bulgaria." *Urban Studies* 41 (13): 2601–19.
https://doi.org/10.1080/0042098042000294583.

Wachsmuth, David, and Alexander Weisler. 2018. "Airbnb and the Rent Gap:
Gentrification Through the Sharing Economy." *Environment and Planning
A* 50 (6): 1147–70. https://doi.org/10.1177/0308518X18778038.

Wilson, Julie, Lluís Garay-Tamajon, and Soledad Morales-Perez. 2022. "Politi-
cising Platform-Mediated Tourism Rentals in the Digital Sphere: Airbnb in
Madrid and Barcelona." *Journal of Sustainable Tourism* 30 (5): 1080–1101.
https://doi.org/10.1080/09669582.2020.1866585.

World Bank. 2017. "Bulgaria - Housing Sector Assessment." http://docume
nts.worldbank.org/curated/en/776551508491315626/Bulgaria-Housing-
sector-assessment-final-report.

Yrigoy, Ismael. 2019. "Rent Gap Reloaded: Airbnb and the Shift from Residential
to Touristic Rental Housing in the Palma Old Quarter in Mallorca, Spain."
Urban Studies 56 (13): 2709–26. https://doi.org/10.1177/004209801880
3261.

Zanini, Sara. 2017. "Tourism Pressures and Depopulation in Cannaregio."
Journal of Cultural Heritage Management and Sustainable Development 7
(2): 164–78. https://doi.org/10.1108/JCHMSD-06-2016-0036.

CHAPTER 4

Airbnb-ed Homes and Everyday Life

Abstract This chapter explores how short-term rental platforms trans-form the intimate spatialities, everyday practices, and the social relations at "home". Drawing on an (auto)ethnographic study of *Airbnb* within the context of Sofia, the chapter shows that the meaning of home is continuously reshaped by the social and emotional relationships that are established between different hosts and different guests. While being disruptive it also argues that the *Airbnb* economy represents an oppor-tunity for some hosts to produce and extract new values from their intimate spatialities and their ordinary practices of homemaking. Finally, the chapter illustrates how, in the process, having guests over oftentimes unsettles and rearranges the social relations between those already living in that "home".

Keywords Home · Commodification · Performance · Autoethnography · Tourism

IMAGINARIES AND SPATIALITIES OF HOME

In marketing campaigns *Airbnb* has often claimed that travellers can expe-rience what it means to "live a local life" by staying in people's homes, and—in so doing—"belong" to distant places (Chesky 2014). By tapping

© The Author(s), under exclusive license to Springer Nature Switzerland AG 2022
M. Roelofsen, *Hospitality, Home and Life in the Platform Economies of Tourism*, https://doi.org/10.1007/978-3-031-04010-8_4

into everyday lives at the level of the household, travellers may supposedly experience "authentic" local cultures away from the "beaten track", signifying a "reflexive and daring traveller identity in opposition to that of the mass tourist" (Gyimóthy 2017, p. 66). As "locals" become the intermediaries who interpret the places they live in for tourists, such tourism experiences ostensibly centre on the relationships between hosts and guests in place (Richards 2017). These narratives suggest that the platform is both enabled by and productive of certain imaginaries and spatialities of home, rooted in an idea of collective global belonging. Such accounts emphasize the normative association of home with positive values, yet it reveals little about how commodification processes may be disruptive of everyday life and contribute to the emergence of entirely new ways of *being at home*. Considering that people's sense of self and "feeling at home" rely on the social and emotional relationships in these spaces (Blunt and Dowling 2006), one may question how home is made and unmade when relative "strangers" move in and out of home with each new transaction. To answer this question, I provide an empirical account of how home is enacted through "home-making" practices in the *Airbnb* economy in Sofia, Bulgaria. Through the perspectives of local *Airbnb* hosts and my own experiences as their guest, I illustrate the diverse and sometimes contested understandings of home and how myriad performances bring home into being. In the paragraphs that follow, I first selectively engage with a literature that discusses the changing role of home in (re-)production of capitalism and late capitalism. After briefly discussing the meaning of home in the sharing economy and tourism, I then expand on the idea of home as being practised and "performed".

SHIFTING MEANINGS OF HOME

Home and Capitalism

In Marxist accounts home has been conceptualized as a site for the reproduction of the social relations that maintain capitalism and of the material bases upon which social life rests (Gregson and Lowe 1995; Blunt and Dowling 2006). Home, here, is described as a site where workers recover in order to continue their work; a place that serves to *maintain* individual productivity under capitalism. In these accounts, the caring and affective labour taking place at home is generally not considered to produce "value" and is therefore accordingly not considered as (waged) work. At the same time, "home" ownership is seen as instrumental to the success of capitalism and to forward an ideological agenda aimed at economic efficiency and growth (Mallett 2004, p. 66). In many Western countries, home-ownership is supported by governments through state policies, and heavily promoted by the real estate industry (Knox and Pinch 2014). It supposedly stimulates workers to remain committed to their jobs in order to pay off their mortgages, and signals identification *with* and the incorporation *of* capitalist values (Harvey 2008). In the last two decades, the global struggles over housing and associated practices of displacement and dispossession are also testimonies of the unmaking of home spaces (Brickell et al. 2017). On-going external pressures such as the affordability of housing, housing instability, and lack of autonomy may also affect whether or not one feels at home (Bate 2018; Lloyd and Vasta 2017). For example, under the pressure of tourist investors who have pushed to convert long-term rental property into short-term rental property, long-term tenants in Barcelona have dealt with expulsions, harassment, rent increase, and affordability problems, as detailed in the previous chapter (Gant 2016).

Feminist scholars have problematized and challenged the earlier conceptual separation of the production and reproduction spheres. The manifold everyday practices people engage in complicate and pervade the supposed borders between "work" and "nonwork" (Mitchell et al. 2012). Moreover, feminist scholars have highlighted the gendered nature of the separation between "work" and "nonwork". Historically, cultures have equated women with home where they serve, maintain and nurture men and children (Young 1997, p. 134). Women remain, until today, the primary "home-makers" around the globe (Bowlby et al. 1997; Duyvendak 2011). The unwaged caring and affective labour that takes place outside the factory walls and offices is in fact value-(re)producing and should be understood as work (McDowell 2004; Mitchell et al. 2012). The household as the primary site where labour power is regenerated is thus "as inextricably bound up with the operations of capitalism as are capital and the state" (Marston 2004, p. 176).

Home in Late Capitalist Economies

Under the impact of globalization and related information and communication technologies, a profound qualitative transformation in the nature, form and organization of labour has taken place (Gill and Pratt 2008).

In late capitalist economies, workers are often expected to be temporally flexible and spatially boundless. By ways of encouraging this work ethos, the spheres of work and home are progressively and oftentimes deliberately defused (see Hochschild 1997 on the decline of "home-at-home"). With the "extensification of work", or, the exporting of work across different spaces and time, people increasingly work from home leading to an "overflow of work into wider social life" (Jarvis and Pratt 2006, p. 338).

Yet another trait of this transformation are working spaces that purposely resemble home and/ or should make workers "feel at home". *Googleplex* (Google's headquarter) for example, incorporates an array of facilities, services and activities previously associated with the private sphere. Here, workers are incentivized to bring their children and pets along to work, get in-office haircuts and massages, use on-site sports facilities, play games, get their clothes dry cleaned and take up classes in personal and spiritual development. While such "home-like" work environments may provide workers the comfort and ease of having certain necessities available within arms-length, another objective of these defused spatialities on part of employers is to incentivize, control and enhance productivity among employees. These transformations engender what Mitchell et al. (2012, p. 3, italics in the original) define as "the interpellation of subjects as *life workers*—the rendering of permanently mobilized bodies in new kinds of technologies of power". The labour performed in late capitalism also increasingly resembles the labour previously associated with the affective and caring spheres of home. The production and delivery of services, information and experiences, through activities such as entertainment, advertising, health care and finance, are realized through labour that produces immaterial products, knowledge, information, ideas, images and affect (Hardt 1999; Hardt and Negri 2000). This form of labour, again, continues to blur distinctions between what is work and what is not, when one works and when not, and where one works and where not, allowing for an identification of both the work sphere and home with economic activities, whether these activities are waged or not. The blurring of these distinctions is further encouraged by current neoliberal agendas in order to produce insecurity in the absence of fixed contracts and the proliferation of cheap unprotected labour. This echoes a "feminization of work" by which workers, male or female, are increasingly precarious and in a constant state of insecurity, and therefore easily subject to exploitation (see e.g. Adkins and Jokinen 2008).

Home in the Sharing Economy and Tourism

The labour that is enabled through platforms like *Airbnb*, appears not much different from, if not an exacerbated form of, the aforementioned precarious labour under late capitalism. What is crucially different, however, is that these platforms explicitly mark out people's private spaces—homes—as sites *for* and *of* production and consumption; a potential income-generating sphere. Botsman and Rogers (2010) in their conceptualization of the sharing economy, optimistically conceive of the home as an "underused good", or, an asset that sits "idle" that could be made available to others who want to use it on a temporary basis. In the "sharing" economies of tourism, home is not merely the *site* where (household) labour is carried out, but in its entire materiality and imaginary is conceived of as a commodity—whether paid for or not. Unlike quasi-commercial and commercial home enterprises such as homestays, guesthouses and bed & breakfasts, both hosts *and* guests are reviewed and rated for their performances in their respective roles. As such, responsibilities *for* and expectations *of* "homely" experiences rest on the shoulders of both "consumer" and "producer" in this sharing economy.

Whereas peer-to-peer accommodation platforms may be a relatively new phenomenon, quasi-commercial and commercial home enterprises such as homestays, guesthouses and bed & breakfasts have long been contentious sites of study (Lynch et al. 2009). In such forms of tourism accommodation, intimacy may also have commercial value. Depending on the demands and behaviour of guests, the hosts constantly (re)draw the lines between the public and private spheres in those sites (Andersson Cederholm and Hultman 2010) while employing different strategies to make guests "feel at home" (Sweeney and Lynch 2007). This work challenges the previous notions of home, in particular those commonly proposed in the tourism literature. In tourism studies, home and tourism have been for long theorized as opposite ontological worlds (Larsen 2008). It is the ordinary and mundane home one escapes from or returns to after having fulfilled one's desires in places elsewhere in the world, through travel. As such, home is treated as a separate sphere, a point of departure or return, or is explicitly left out in the conceptualization of social life (ibid.).

In an age of high mobility and changing local contexts, the realities and practices that constitute home in the twenty-first century have ostensibly changed. They have arguably spurred an increased sense of translocal- or transnational belonging (Lloyd and Vasta 2017). Tourism enables "different modes of attachment to and detachment from places, cultures and people; different modes of belonging in the world; hence we may conceive of tourism as a way of positioning oneself in the world" (Haldrup 2009, p. 54). Tourist performances and experiences, Haldrup (ibid.) argues, are also an important vehicle for the emergence of cosmopolitan orientations in everyday life.

In the past decade, platforms like *Airbnb* and *Couchsurfing*, have (quite successfully) tapped into a cosmopolitan fantasy of "being at home in the world" (see Germann Molz 2007). Through social networking technologies, such platforms have connected millions of strangers all over the globe in their desires for more individuated encounters. By enabling peer-produced-hospitality in everyday and homely environments, these platforms ostensibly centre on experiences in rather than of place (Russo and Richards 2016). As such, they have also inspired new forms of sociality (e.g., Germann Molz 2013, 2014) and have refuelled debates on cosmopolitanism (Picard and Buchberger 2013; Zuev 2013). These accounts also suggest that travel may open up new territories where people feel at home. In an analysis of accounts of round-the-world travellers, Germann Molz (2008) contends that—in their privileged positions—travellers may also feel at home *in mobility* (Ellingsen and Hidle 2013). By engaging in "small acts, embodied practices, and familiar routines" travellers make themselves at home "online, on the road, and in the world as a whole" (Germann Molz 2008, p. 337). The sense of "being at home", Accarigi (2017, p. 192) argues, can be disassociated from a geographical location and "replaced by belonging through specific everyday practices".

To take on a perspective of home as "practiced" or "made" rather than to "*be* at home", Lloyd and Vasta (2017, p. 4) argue, offers "a new set of possibilities to make ourselves at home in relation to others". However, home is not merely "made" through the interrelation of social relationships, spaces, and materialities. Home may also be understood through "*unmaking*", a "precarious process by which material and/or imaginary components of home are unintentionally or deliberately, temporarily or permanently, divested, damaged or even destroyed" (Baxter and Brickell 2014, p. 134; but also, Brickell 2014). What is said, done and "acted-out"—or, performed—transforms spaces and places into what they come to be or what we want them to become (Gregson and Rose 2000). These performances are embodied and give way to a corporeal approach to an understanding of home in tourism's sharing economy. In the following paragraph, I briefly expand on the performative approach that I have taken in this research, and how consequently this approach has influenced my choice to employ a set of qualitative methods.

Approaching the Home Through Performance

The arguments presented in the following paragraphs are informed by the theories of performance and the role of performance in bringing social life into being. Here, I rely predominantly on the work of Erving Goffman and Judith Butler, recognized as two of the most influential "performance" theorists in the geographical literature (Gregson and Rose 2000). While their conceptualizations of performance differ, aspects of both conceptualizations have been key in the interpretation of home in this research. Firstly, I rely on Goffman's (1956) dramaturgical approach, which argues that social life may be thought of as "staged" by conscious actors who perform for an audience according to the scripts and codes of conduct. By distinguishing between a front- and backstage, Goffman suggests that the self performs various roles according to the different stages one finds oneself on. The "frontstage" signifying those spaces (un) intentionally employed for (routine) performances that require a certain setting, décor, and/or appearance. The "backstage", on the other hand, signifying those spaces that are relative to the given performance; reserved for preparation, practice, recovery and retreat. I employ Goffman's notion of performance to analyze the possible structural divisions of social establishments within the context of the *Airbnb* "home". In particular, I examine how home (both materially and symbolically) is produced, maintained and regulated as a "tourist stage" (MacCannell 2013).

Secondly, I employ Butler's (1986, 1988, 1990, 2004) conceptualization of performance. In Butler's theory of performance, which refers predominantly to the questions of sexuality and gender, identity is not so much a construct; rather, it relates to what we do or what we "perform" (Butler 2004). As such, performance is imbued with an ability to destabilize and transform social identities that are all too often considered stable. Like Gofmann's "codes of conduct", Butler argues that norms serve as guidelines in social intercourse throughout life, consequently guiding one's performances. For norms to come into effect, embodied practices need to be repeated. However, the repetitions of embodied practices are never entirely the same; "bodies never quite comply with the norms by which their materialization is impelled" (ibid., xii). In fact, they are *reinterpretations* of what is assumed must be performed, calling the absolute force of the norm into question. Moreover, performances, can also offer "an opportunity to mark subjectivity" by rebelling against the codes of conducts, scripts and norms (Edensor 2001, p. 75). In the work of Butler, the notion of performance thus allows thinking beyond what one *is*, and focuses instead on what one *becomes*. Employing Butler's notion of performance allows considering those performances that deviate from normative conceptualizations of home.

AN (AUTO)ETHNOGRAPHY OF HOME

The present study relies (in part) on autoethnography. In performing autoethnography, the living and embodied subjective self of the researcher is considered an active agent and a constitutive part of the research process; the researcher is the "epistemological and ontological nexus upon which the research process turns" (Spry 2001, p. 711). In my capacity as a guest, I thus considered myself an active agent in *making*-home rather than a passive observer who sits on the receiving end of the *Airbnb* experience. An autoethnographic approach allowed me to focus on my embodied and affectual encounters in place. As a researcher-guest, I self-witnessed and committed my body to the intimacies of experiential encounters. In doing so, Caroline Scarles argues, we become witnesses to a place for ourselves (2010, p. 911). As a relational approach, autoethnography enables a variety of ways to engage with the self in relation to others, to culture, to politics, and other aspects (Allen-Collinson 2013).

My autoethnographic accounts of the day were either recorded in writing (in a diary) or on video (via my smart phone). These accounts reflected the practices and interactions that made an impression on me during my stay. At the same time, I took note of the materiality and sensory richness of the home as these played a crucial role in enacting interactions (Haldrup and Larsen 2006; Haldrup 2017). I took care to observe how things relate and through what practices, rather than what they essentially are, in a static form. Adopting an autoethnographic approach allowed me to directly explore how our "homely" performances, like touristic performances, "involve, and are made possible and pleasurable by, objects, machines and technologies" (Haldrup and Larsen 2006, p. 276, but also Accarigi 2017). Analyzing my diary entries and videos eventually allowed me to trace how "feeling at home" shifted according to an amalgam of materialities, experiences and embodied practices as well as the changing relationship to my hosts over time.

To study how home was (un)made through the performances of both the host and guest, I stayed in 11 different *Airbnb* homes in Sofia, Bulgaria for respectively two to five nights each. During these stays, I was hosted by Anne, Nico, Eva, Lisa, Pia, Jeny, Maria, Tina, Adrian, Ria, Iordan, and Vera. In-depth interviews were held with each of these hosts and I refer to these interviews directly and indirectly in this chapter. The interviews took place according to the hosts' preferences. On most occasions this meant at home, while several other interviews were conducted in bars and restaurants in Sofia. The interviews were taken individually and lasted between 15 minutes to approximately 2.5 hours; however, casual conversations that also informed my research took place throughout my stay and were taken note of in a notebook. All interviews were taped with the hosts' consent with the exception of Ria's interview. Prior to this interview Ria expressed her worries for possibly being exposed and sanctioned by the authorities and hotel industry for using the *Airbnb* platform. Ria asked me to take notes during the interview instead. The interviews were transcribed verbatim and, together with other data, were analyzed through qualitative data analysis software MAXQDA. The textual accounts of the interviews were searched for common themes related to the research objective. In the following paragraphs, I have used pseudonyms to guarantee my hosts' anonymity.

Selecting Airbnb *Homes in Sofia, Bulgaria*

The hosts and their homes were first and foremost selected based on their set overnight rates (between €10 and €20) as I relied on a limited budget. Secondly, hosts who received ratings and reviews indicating recurring unacceptable behaviour or homes that were considered unfit by previous guests were not taken into consideration for safety reasons. An additional condition in the selection process was that the homes had to be available for the requested days during the months in which the "fieldwork" took place.

As I took on an active participant role, my hosts were provided with a full explanation of my position as a researcher and what my research was about. When making booking inquiries, I informed the hosts that my stay in Sofia was tied up to my research and asked if they were interested in sharing their experiences with me. The large majority replied positively to my inquiries whether or not they were able to host me. When I stayed in Sofia, I had already been using *Airbnb* for personal travel for over a year. In consolidating my booking requests in Sofia I was thus able to rely on numerous positive reviews and ratings that I received from my previous *Airbnb* hosts. This form of "digital social capital" has arguably aided me in getting access to the "field"; moreover, the positive reviews that I received after each new stay in Sofia arguably aided my access to other *Airbnb* homes.

During these 11 stays, nine out of 12 self-identified "hosts" were present on the premises most of the time throughout my stay. During the other stays my hosts sometimes spent time with me during the day or evening, but slept elsewhere (referred to as Pia, Maria and Ria in the findings). During five of my stays, other people (tenants, family members, partners and friends) were sleeping on the premises too but rarely engaged with me throughout my stay. Some introduced themselves to me but took little or no responsibility in the hosting activities. With few exceptions, my hosts were white, middle-class highly educated Bulgarians who resided in Sofia and used the platform for their own travel endeavours. Except for Anne, all of my hosts spoke English, were (highly) mobile and to (some degree) open to cultural differences. They also used the platform to *increase* their earnings rather than relying on *Airbnb* for a sole income. This attests to other studies on the sharing economies of tourism as exclusive marketplaces for cosmopolitan citizens; a community marked by privileges such as the political right to mobility and the means to travel as a requirement to "be at home in the world" (see Germann Molz 2007, 2008; Picard and Buchberger 2013; Zuev 2013; Schor 2017).

Unlike other cities in Europe, the municipality of Sofia had not yet imposed specific rules and regulations on the *Airbnb* platform and its users at the time I conducted this research. Short-term rental platforms were not treated any differently from other commercial short-term rental agencies. However, under the Bulgaria Tourism Law, homeowners were officially obliged to register their apartments with the municipality in order to obtain a licence to rent them out to tourists (Baltova and Vutsova 2021). Although no official numbers exist, it has been speculated that a large majority of the hosts did not possess such a licence at the time (Neychev 2017).

Researching the Airbnb Home

While this study is situated in academic literature focussed on the global political economy of *Airbnb*, the aim of this (auto)ethnography is not to generalize or to explain the experiences of a wider universe of *Airbnb* hosts. Rather, it depicts a series of case studies, of homes and encounters between myself and the hosts, each detailing the specific culture and micropolitics of the related context (see Mohanty 2003).

It is important to state that being a white, able-bodied, middle-class woman with a privileged Northern European upbringing makes my knowledge necessarily partial, contingent, and *situated* (Haraway 1991). On many occasions I found myself wondering what gave me the right to write *about* the lives and homes of Eastern European "others"? There is of course no straight-forward or easy answer to this question. I have, however, attempted to closely listen to my participants and tried to detail their stories, worldviews and social realities in a way that hopefully shows their inherent multiplicity and plurality.

My positionality has indeed also affected my access to the field. Some of my hosts expressed that they were positively inclined to accept my booking request based on their judgement about my online *Airbnb* profile. This profile included a picture and details on my profession, education and other biographic details. Several hosts explicitly mentioned that they only accepted booking requests from people who reflected their preferences in terms of gender, class, age, sexuality, religion and race (see also Karlsson et al. 2017). Ria, for example, declared that she would not accept bookings from non-white and/or Muslim guests. Various cases in my research confirmed the explicit and implicit practices of discrimination that occur on the platform (as shown in Edelman et al. 2017). In other words, the processes of inclusions and exclusions shape *Airbnb* homes—like they do other non-commoditized homes (see also Blunt and Varley 2004).

Setting the "Stage": Material Accounts of (Post-)Socialist Homes

With some exceptions, most of my hosts live in a one-family (high-rise) apartment, which were built between the 1960s and 1980s. Bearing concrete constructions, these apartments were marked by a single coherent architectural style, a result of the socialist regime's embrace of a specific strand of modernist architecture in a context based on anti-capitalist and anti-bourgeoisie ideologies and driven by the political objective to provide everyone with a home (Hirt 2006). The materialities of these homes carry with them significant historical and political meaning. At the same time, and to speak with Haldrup and Larsen (2006), such materialities crucially condition the practices of homemaking and the feelings of homeliness in several ways even today. By fleshing out some of these "inherited" socialist materialities I firstly show how specific objects and home spatialities both enable, constrain and guide performances of *Airbnb* homes in Sofia. Secondly, I reflect on how some of these socialist material legacies have become part of new touristic values and are purposely marketed as such.

Energy Monopolies and the Visceral Home

Most of the apartments I stayed in are still powered by centralized heating and power systems that were initially installed and operated under the socialist regime. Since Bulgaria's transition to a so-called "free market regime" and to liberal democracy, these systems have been predominantly operated and owned by private (foreign) companies after the state sold off its majority stake, creating true energy monopolies in this way (Minchev 2013). These companies are occasionally termed "the mafia" in the accounts of my hosts; a name arguably earned for their practice of charging Sofia's residents constantly fluctuating and highly inflated fees while providing unreliable services. High electricity bills, together with low wages and pervasive employment insecurity, have been at the root of the nationwide anti-government protests in 2013 (Tsolova 2013; Koycheva 2016); protests that several of my hosts have actively participated in (and mentioned in the interviews). Two years after the revolt in 2013, the owners of these centrally operated heating systems still shape how homes are felt, lived and experienced in Sofia. On numerous occasions, and for extensive periods of time, there was no central heating flowing into the homes where I was staying, bringing indoor temperatures closer to outdoor temperatures, which averaged between 0 and + 15 degrees Celsius at that time of year. The enduring cold that I experienced in some of these homes came with a constant sense of unease and eventually brought about illness on my part.

Home, throughout my *Airbnb* experience in Sofia, inadvertently became a profound affective, sensuous and visceral experience; an experience that made me long to be "elsewhere" on many occasions. It led to the enduring feelings of *non-belonging*, paradoxically running contrary to what the *Airbnb* commercials would like guests to believe. As the cold conditioned and significantly restricted my embodied practices, I found myself leaving the apartments in search for better-heated public spaces during the day. Consequently, the time to form (potentially) affective relations of belonging with my hosts became limited as I stayed away from their homes until the late evenings. Once back, I would promptly retreat to the solitude of the bedroom where I could stay under the covers fully dressed. This echoes Haldrup's (2017, p. 53) account of the important role materialities play in enacting relationships between people, as well as "reinforcing bonds as well as boundaries between the home and the world outside". Although I tried to make up for my absence by inviting my hosts along for walks and meals outside, my alienating habits eventually left me feeling inadequate as a guest. A guest who—in the political economy of *Airbnb*—also gets rewarded (or disciplined) for their (un)homely performances through review and rating systems, and the algorithms that the platform operates to ensure "quality control" (see Roelofsen and Minca 2018, for a critical discussion of these systems of measurement).

The constant cold due to the lack of adequate heating systems thus became a mere obstacle to me *feeling at home*. It also showed that the materiality of the intimate and personal spaces of home in Sofia is deeply interwoven with wider national politics and power relations. Home, indeed, was not "a secluded 'private' space but a space in which outside forces make their entry" (Haldrup 2017, p. 53). The materialities that underlie post-socialist housing in Sofia came to determine the conditions by which bodies are both enabled and restricted to engage in the ordinary practices of homemaking, particularly during the colder months of the year. Several hosts shared with me during their stay how they dealt with the insecurity of not being able to pay their energy bills and consequently being unsure if they could provide their guests with the warm and "homely" environment they might be used to in their own homes. However, at the same time, an additional *Airbnb* income enabled them to continue paying some of their household bills, thus appropriating the platform to fulfil their own homely needs. Although this allusive interdependence on the platform was never easy, it also opened up a space for alternatives.

Home Heritage

Home and place identities come about through the creative expressions of hosts and their display of symbols and artefacts in hospitality practices (see also Di Domenico and Lynch 2007). Pia, one of my hosts, shared with me how she created value out of the historicity and materialities that are so specific to the multiple homes she puts on display through *Airbnb* platform. As an intermediary, Pia takes on the work of representing several *Airbnb* homeowners in Sofia, receiving a small commission for welcoming guests and being available to assist them throughout their stay. The home-owners she represents either only speak Bulgarian or do not have the time nor interest to engage in the hosting labour but are interested in earning the related income. These particular homeowners do not live in their homes permanently when guests are present. They often outsource the labour of cleaning their homes to others, making their presence around the home limited or non-existent.

Although Pia does not spend a lot of time with guests at these homes during their stay, she takes on an important role in framing and (re-) distributing socialist values through the homes under her umbrella on the *Airbnb* platform. Pia argued that she has perceived among *Airbnb* guests a certain interest for everyday life under socialism, a desire for the supposed "authentic" socialist Other and their spaces. Pia's observation mirrors MacCannell's (2013) critical account of tourism, according to which the tourist alienation of their own life has led them to search for "reality" and "authenticity" in the purer lifestyles of other cultures.

The tourist, here, can be conceived as an emblematic modern subject that seeks to escape the humdrum of ordinary life and attempts to find "an outlet for existential anxiety over the precariousness of the modern condi-tion" in tourism (Minca and Oakes 2012, p. 6). However, such desires are often not well understood by Pia's clients who actually used to live such lives under socialism in the homes they now rent out. They find it hard to think of marketing an essentialized socialist identity and culture through their homes on the platform. The materiality of home evoked involun-tary memories of otherness (Morgan and Pritchard 2005, 42); a socialist past they no longer identified with and tried to dissociate from (see also Kaneva and Popescu 2011). Notwithstanding, Pia shared with me that she continues to recommend her "clients" to conserve their homes in "original" and uniform socialist character and refrain from polishing the interiors into a "uniform" globalized and capitalist "IKEA jacket". Pia suggests to the homeowners not to replace their domestic material objects acquired under socialism with "modern" ones, as socialist life and decor has recovered its value under tourism. They had the potential to authenticate the guest's experience. What her clients consider "bad taste" or backward—furniture, household equipment and technologies, and decorative items from the socialist era—was picked up by Pia as potentially value-generating. In their absence at home, such items referred to the life histories of their owners, and, like souvenirs, had "the effect of bringing the past into the present and making past experience live" (Morgan and Pritchard 2005, p. 41). Such artefacts had the potential to contribute to a lived experience of socialism, a multi-sensory experience by which home could be seen, felt, heard, and smelled.

This gives way to thinking of the *Airbnb* home as an "exhibit of itself". A home as "heritage" by which some hosts or host-representatives like Pia give their "ways of life" and homes a second life (Kirshenblatt-Gimblett 1998, 150). As certain materialities and historicities have different meaning and significance under tourist consumption, it goes to show that they can also give renewed purpose to homes that were once conceived as less valuable and in need of improvement. Being perceived as "exotic", everyday objects connected the contexts of "home" and "away" into a setting for the tourist performances (Haldrup 2009, p. 55).

Pia and the other homeowners feel they exert a certain level of control over the ways in which their homes get commodified through *Airbnb*—together they determine themselves the extent and ways in which their homes get absorbed in the platform economy and for what they put it to use. This echoes Mary Louise Pratt's (2008) concept of the *contact zone*, meaning the spaces where cultures intersect and grapple with each other. Emerging from this zone is the phenomenon of "transculturation", which challenges viewing encounters from one cultural perspective or from opposition (ibid.). Rather than assuming it is the *Airbnb* guest who "consumes" the culture and homes of their host during her visit, some hosts purposively select those materials and practices that are deemed valuable by their visitors. Employing the contact zone as a tool then becomes a more affirmative way of describing the complex transformation of *Airbnb* homes.

Sharing Home with Strangers

In the *Airbnb* experience, everyday practices that are usually done alone or between people in an intimate relationship are given value by practising them with strangers "in the home". They include socializing and eating together, commuting, taking a shower, walking around in pyjamas, relaxing on the couch, and sleeping, among other activities. This suggests that contemporary tourism (and society more broadly) increasingly relies on affective capacities and practices marking out people's private spheres (Veijola et al. 2014). Having strangers under the same roof also gave my hosts a different lived experience of home, which positioned them differently in relation to their home. Eva shares:

> In a way, [*Airbnb*] is always a dangerous thing. That you go to someone's house and you live there, you sleep there. At the same time, I am hosting somebody I do not know. You will use my apartment, my flat, my toilet, my everything. I am leaving all my stuff there, my expensive belongings. There is always this dangerous part in the whole experience. [And] many times the guest is too cold [referring to their ways of interacting]. You will be like: 'This is your key, if you need something this is the map, this is where I have my coffee, this is a good restaurant. Thank you. This is it.' And sometimes it is totally the opposite: 'Come have breakfast with me, tell me something, walk with me while I go to work.' And the experience is totally different because [that person] changes your entire experience.

Sharing home with strangers, as Eva's account suggests, is a symbiotic process of unmaking and remaking home (Baxter and Brickell 2014, p. 135). It changes along with the different social and emotional relationships that are formed between hosts and guests. What Eva's account also points out, is that sharing home may challenge home as a source of "ontological security" (Dupuis and Thorns 1998) and feeling ontologically safe. In the *Airbnb* experience, where lockable bedroom doors often lack, sleeping is drawn into a sticky negotiation with every new stay. Veijola and Valtonen (2007, 23–24) in their analysis of sleep in tourism, contend that sleeping, today, is one of the most private and intimate acts: "you either sleep alone or with someone you know well, [the] dormant body is not visible to strangers".

My sense of home and ontological safety was severely disrupted when I rented Maria's entire home. I had rented it in its entirety precisely to get a break from sleeping in the presence of strangers. Hours after having checked into Maria's apartment somebody repeatedly and forze-fully knocked on the front door. After a minute of being startled and not knowing what to do, I finally opened the door. An intoxicated man intro-duced himself as Maria's husband. He was the owner of the apartment and demonstrated a set of keys in his hands. Without asking permis-sion to come in, Maria's husband passed me in the hallway, and made himself comfortable at the kitchen table. In the minutes that followed he subjected me to an interview about my stay while giving me a slurred speech about the state of tourism in Bulgaria. He had been informed about my position as a tourism researcher by Maria, which I had also detailed in my online *Airbnb* profile. After half an hour, he left, and I realized he had trespassed what was at least temporarily "my" home. I was left in a considerable state of uncertainty as his set of keys allowed him or anybody else to enter the apartment at any given time. My imag-ined sense of privacy and safety in that space had been wiped out in an instance. As my corporeal vulnerability was brought to "the forefront of my consciousness" along came several embodied consequences (Allen-Collinson 2013, p. 295). In the days to follow, I had my ears on full alert to listen if anybody was wandering around the front door. At night I left the lights on so I could see better if anybody came in. After two consec-utive sleepless nights I decided it was time to move out—this home had been forcefully unmade.

Airbnb homes in both my own experience as well as my hosts' expe-rience thus became ambiguous sites of, at the same time, belonging and non-belonging. Places where various practices of trust-making were *by default* part of the "homely" experience and the mutual endeavour of "making-home". During my *Airbnb* stays with hosts present, such trust consolidated around an understanding and respecting of each other's emotional and bodily needs. In doing so, I continuously tuned into my hosts' daily intimate routines and to what I perceived as their "hygienic standards". I would make sure not to come home after 11 pm; not to use the shower in the morning before they did; or, waking up and moving around the living room and kitchen not too early but also not too late. Along with becoming overly concerned with keeping my bedroom in order (by making my bed every day and keeping my clothes and personal belongings stowed away), I disciplined and controlled my body in order to meet, what I thought, were socially accepted standards of being a "clean guest". I was careful not to make any loud or awkward sounds in any spaces of the home. Each time I visited the bathroom, I double-checked if I did not leave any bodily products behind that might socially or culturally be deemed as "dirt" or "waste" (Isaksen 2002). Through these embodied practices, I aimed to avoid provoking disgust in my hosts. Such practices ostensibly subscribed to my own ideas of "appropriateness", and to "culturally-validated values of 'proper' bodies and 'proper' femininity" (Fahs 2017, p. 192).

Besides relying on my social skills in bonding with my hosts, I thus engaged in various forms of affective and embodied labour in a greater endeavour of making *each other* "feel at home" throughout our stay. During everyday encounters with my hosts, I continuously *smiled*— an "embodied display and an act of amiable hospitality" (Veijola and Valtonen 2007, pp. 20–21). And in being overtly but *not* insincerely appreciative for things like a clean towel and a spontaneously offered cup of tea, I attempted to operate on my hosts' bodies as to "provoke a state beyond what can be cognitively communicated" (Dowling 2012, p. 113). Paradoxically, such practices ran counter to what I associated with *being at home*-at-home; a place where I could be emotionally unrestrained and enjoyed a (literally) embodied freedom. A place where I could be grumpy, loud, late, early, messy and above all *dirty* (Veijola et al. 2014, p. 1). Instead, I fully engaged in co-providing my hosts with a clean home as a mutual act of hospitality and a way to slowly rid my "strangeness" from their own home.

In a home that was not even mine, I "stage-managed", ordered, and disciplined myself, intuitively tuning in to the rhythm of my hosts' day. A tiring process that, in producing fatigue, drew me even further away from "home" but at the same time felt like a more ethical way to relate to these unknown others. Contrary to what Veijola and colleagues (2014, p. 3) suggest, stranger/guests may no longer be as messy as they have been theorized, especially in a reputation economy like *Airbnb* where hosting and guesting bodies are monitored, controlled and disciplined through its ranking modalities (Roelofsen and Minca 2018). Instead, they may have become increasingly complicit in reproducing that same old "tidiness" paradigm that tourism theorists, planners, activists and tourists have become so fixated on.

BORDERING THE PRIVATE
AND INTIMATE SPATIALITIES OF HOME

After our brief introduction, Anne instantly leads the tour of the two-bedroom apartment as part of her performance as host. The tour proceeds in moderate silence, as we both seem uneasy to express ourselves in a language we do not control very well. Anne leads, entering the different doors in the house. She first guides me to her son Nico's former bedroom, which is adjacent to her own bedroom and has access to a balcony facing the inner courtyard, which is now covered in snow. This is where I will sleep the next couple of nights. The 15-square metre room features a double bed, a desk, a chair and a couple of empty cupboards. Anne continues the tour through the kitchen and living room and shows me the workings of the toilet, which has some particularities to it. As she concludes the tour in the hallway, it becomes apparent that her bedroom is the only one that she will not show to me: the place where her body lays to rest will not be on display today.

In my *Airbnb stays*, "home-tours" like Anne's were commonplace and served as an implicit bordering practice (see also Andersson Cederholm and Hultman 2010) to negotiate the separation—or lack of separation—between shared private spatialities (e.g., the living room and kitchen) and the remaining intimate spatialities (e.g., the bedroom). In a similar vein, when Eva and Lisa toured me around their apartments, they made clear that I would be sleeping in their bedrooms while they would be sleeping on the sofa bed in the living room throughout my stay. In an interview Lisa recalls the importance of her bordering practice, "I *try* not to bother at all the people who stay with me. But I am always saying 'This is it – [pointing at the couch] – I will also stay *here*' so the guests have it in mind".

In this respect, Goffman's conceptualizations of the frontstage/backstage interplay in tourism seem to be implicitly operationalized in the first encounters with my hosts. However, while there seems to be a common understanding of "back-staging"—that is, defining spaces in the house where both my hosts and I would be able to retreat and rest—the act of deliberately delineating the "backstage" on part of the host implied that certain borders were already transgressed by the *Airbnb* guest's presence and/or needed to be respected by the guest's presence. What this shows, I would like to suggest, is that Goffman's sharp separation between front- and backstage in the framing of social relations is here put into question by the "literally" embodied negotiations between host and guest, and the related establishment of constantly blurred borderings in the home context.

The experiences I had during my fieldwork are perhaps more in line with a Butlerian understanding about the embodied performance of home. The presumed and/or *aspired* radical separation between front- and backstage was, on the one hand, endlessly challenged by the constant possibility of trespassing the related (in)visible borders making the "*Airbnb*-ed" spatialities of home. On the other hand, the hosts openly discussed the existence of a presumed backstage that the guest should not access.

These very spatial practices reflect in many ways the fluidity of the literally embodied experience of hosts hosting in their own home. Such fluidity and complexities clearly emerged on multiple occasions and in multiple forms during my fieldwork. For example, host Iordan set out to literally and figuratively "erase" himself/his body from his apartment by retreating to his own bedroom throughout most of his guests' stays. He would only make use of the newly constituted "common" spaces when he was sure the guests had left the apartment. Iordan explains he does so in order "to leave the most possible space to my guests", implying his own presence would stand in his guests' way of feeling at home. Iordan is not alone in performing such empty and silenced spatialities; when *Airbnb* guests stay over, Anne and Nico discipline themselves by "trying to be calmer", and try not to argue with each other in order "not to disturb guests". In a similar vein Jeny contends,

> [When hosting] I leave more time [in my daily routine] to interact with [guests] and not to somehow interrupt them when using the kitchen or the bathroom. I try to get into the rhythm of the guests somehow [...]. Also the whole family, like, we are coordinating our actions more, like, when everyone is going out and coming back. If the guests want something, this means somebody has to be [at home], always.

These practices give to think about the complexities inherent to the promoted and displayed intimacy that is produced by the platform's core travel philosophy. At the same time, the coordinated practices of care and hospitality hint at changing dynamics between the people who make home, and the changing relationship between them.

Changing Relationalities of Home

[In the beginning] my girlfriend did not feel safe in the house while we were hosting beautiful women. (Adrian)

[Hosting] empowers my mother, because she gets to see the world just by sitting in her room. *Airbnb* is empowering, this [economy] wouldn't be possible twenty years ago. Just like with one click creating a... I don't know... put some photos online and people come here from everywhere. So yeah, I think it is very powerful and this experience is changing her life. Because she is getting in touch with different cultures, different people. (Nico)

The unfixed and dynamic nature of home is reflected in the two different accounts of *Airbnb* hosts Adrian and Nico. When relative strangers temporarily move in, it is not just the dynamic between hosts and guests that challenge the meaning of home, similarly those already "at home" may start relating differently to one another *and* to their homes. Adrian had to "work hard" to convince his partner that these "beautiful" *Airbnb* guests who had entered the safe haven of their relationship, *their home*, posed no threat to its fidelity. This particular case illustrates that ideas of home are defined in comparison with what is *not* a home: a place of sexual promiscuity or infidelity. It attests of a relational understanding of home, in which material and imaginative geographies of home are forged in relation to "foreign" or "unhomely" practices (Blunt and Dowling 2006, p. 142). Adrian's account further reveals a normative understanding of home as a heterosexual place that facilitates everyday heterosexual practices and intimacies (e.g., Gorman-Murray 2006; Johnston and Longhurst 2009, chapter 3; Morrison 2012). Home, here, is supposedly a monogamously "safe" location that is unsettled by the arrival of the sexualized and supposedly "predatory" stranger-guests.

The presence of stranger-guests, however, can similarly destabilize home as a space of alienation and oppression, which is mirrored in Nico's account. While home in early feminist accounts has been described as a place that removes women from politics and business (see Blunt and Dowling 2006), the *Airbnb*-home-as-work-and-travel-destination also empowered Anne and brought renewed self-fulfilment. Throughout her adult life, Anne had expressed her sense of self as a mother through home-making practices in her apartment. After Nico left home to work abroad—and Anne was consequently left with Nico's empty room—it was Nico (a fervent *Airbnb*-traveller himself) who proposed to rent out his old bedroom on the *Airbnb* platform. In this way, the empty bedroom could temporarily bring new life to their home and the derived income could cover the basic costs of living of Anne, who had been unemployed for some time.

With Nico's bedroom taking on a different purpose and becoming imbued with monetary and symbolic values on the *Airbnb* platform, Anne began feeling validated for her home-making practices that had sometimes gone unrewarded or taken for granted throughout her life. Anne has progressively started to relate differently to her home and Nico's empty bedroom, which throughout her life had been marked by her role as mother and home-maker. She also began to take on a different sense of herself as an entrepreneur in this newfound economy and came to identify herself as a knowledgeable "local" vis-à-vis the *Airbnb* guests. This echoes that the relation between individual subjects and places gives places, like home, meaning and their identity (Massey 1991, 1994).

In analyzing Nico's and Anne's account, I also recalled Massey's progressive sense of place (1991, p. 25) by giving account of the shift in the power-geometry that began to take place in Anne's home. The initial social relations that made up home before Nico left the house to work abroad began to alter and changed what home came to mean. Besides being mother and son, Anne and Nico have now also become "professional" partners and both take up different roles in commodifying their home through *Airbnb*. This is to say that the new sharing economy is not merely about the experiences and the evolving relations between hosts and guests: with renting out home come new divisions of labour, new value-generating activities and materialities, and revalorization of social relations that have long existed within the household.

While the increasing presence of "others" may challenge a comfortable sense of belonging (Lloyd and Vasta 2017, p. 1), as was the case with Adrian and his partner, it may also challenge the naturalized meanings of home as a place of constraint and oppression. Before starting their *Airbnb*, Anne's home-making practices never really put her much on the receiving end of the values those practices generated. The beds she made, the meals she cooked, the bathroom she cleaned, these were all naturalized and normalized practices associated with her role as a mother. Now, her guests oftentimes showed appreciation and paid her for her efforts and explicitly acknowledged her efforts in written testimonies online.

In their newfound economy, Nico and Anne made an arrangement on how to organize and divide the labour it entails to *make home* for their guests; an arrangement that is equally gendered. Nico takes care of the written online communication with (potential) guests. He "pre-selects" the guests he would like to stay with his mother by checking their credentials. He takes on bookings, manages the accounting, and does "quality-control" by monitoring and relaying the reviews and ratings of Anne's performances as a host. Nico carries out the administrative and managerial work behind his desk, somewhere away from home. Anne, on the other hand, takes on the preparatory work, cleans the rooms, welcomes the guests upon arrival and carries out the emotional and caring labour when guests are around. She also advises her guests on which places to visit when they are in Sofia, if they speak the same language that is. After transactions have taken place, it is Anne who receives the monetary compensation for their joint efforts.

Importantly, while the commodification of home may (re)valorize and (re)signify home-making practices, it does not necessarily eliminate the presence of the broader gender inequality in the *Airbnb* household, especially considering the aforementioned division of labour between Nico and Anne. Like Meah and Jackson (2013, p. 592) in their study on middle-class households, making-home in the *Airbnb* economy is mostly a *lifestyle choice* for the men whom I interviewed and stayed with during my fieldwork. The "new and cool" rhetoric espoused by the technologically advanced *Airbnb* platform has, to some extent, de-stigmatized domestic chores both as low-status work (Schor 2017, p. 275) as well as inherently feminine work. However, several hosts expressed that household chores and the work associated with hosting are still seen as a "nuisance" and in some cases are therefore outsourced to "other" women. Future studies may thus consider shedding light on the possibly gendered, racialized and classed labour that currently takes place in the sharing economies of tourism and tell us more about the power relations that these economies bring forth, or, perpetuate.

CONCLUSION

In this chapter, I have reflected on the situated and embodied everyday performances of home taking place within the *Airbnb* context. While the platform seemingly rotates around a relatively essentialized idea of home as a place of "belonging", this chapter has hopefully demonstrated the inherent contested nature of home in the sharing economy. Like Lloyd and Vasta (2017), who propose to move away from home as a static location and instead employ an understanding of home as "practiced", I have suggested to explore home as "performed". Drawing on Butler's conceptualization of performance, I have tried to show that while home is often considered a stable construct, home is continuously "done" or "undone" in relation to something or someone other (Butler 2004). Foregrounding performance as an analytical approach, this chapter has attempted to stimulate thinking beyond what home *is*, and to reflect instead on how home *becomes* (un)homely (Blunt and Dowling 2006, p. 26).

In the case discussed here, the political and historical specificities, as well as the materialities of the visited homes have significantly shaped the ways in which the ordinary practices of homemaking unfolded. During my fieldwork, the materialities of home shaped the conditions that led to a deeply embodied sensation of "unhomeliness" on my part—while, at the same time, strongly shaped my appreciation of the hosts' everyday lives. Similarly, a variety of apparently trivial and undervalued objects in the home were (re-)engaged with in order to *stage*, in the words of Goffman (1956) and MacCannell (2013), an experience of "real life" under socialism. These objects thus imbued significant use-value *as well as* a symbolic value in the performance of home (Haldrup and Larsen 2006): while satisfying the guests' desires for an "authentic" experience of home, the materialities of home were also crucial in making homely geographies performable.

Throughout my stays, a complex mix of corporeal and affective practices continuously reconstituted home as an intimate spatiality; one in which it was never quite clear who was the guest and who was the host. The performances of *Airbnb* homes in the cases exemplified capture the ambiguity of the often taken for granted opposition between host and guest. Moreover, such performances challenge home as a place of belonging *or* non-belonging (Blunt and Dowling 2006, p. 255). Some of my hosts, for instance, admitted being self-disciplined into silent bodies or deliberately discounted their own intimate practices around the house to avoid the disruption of their guests' sense of home. By means of "giving more space to the guests" some of my hosts went so far as to (temporarily) "erase" themselves from the shared spaces of their apartments and provided the visitors with a sense of being in a "private" home of their own. Other hosts (implicitly) bordered certain intimate spaces, such as their bedroom, or, alternatively shared certain intimate practices only with some guests, but not with others. By performing imaginary "insides" and "outsides" within the home, they attempted to maintain or protect their own sense of "homeliness" while having guests around. Similarly, I engaged in various embodied practices that I sensed pertained to the norms of "appropriateness" and "hygiene" in those specific contexts. As such, home does not come into being through one singular act but through reiterative and citational practices (Butler 2011) that are guided by normative understandings of home.

Finally, what my experience in Sofia has shown is that the homes involved in the *Airbnb* platform economy represented a key site through which social relations took on a new direction and meaning. Importantly, this chapter has proposed to move away from the emphasis on encounters and relations between "strangers" (e.g., Bialski 2012; Germann Molz 2007; 2014; Picard and Buchberger 2013) and to detail instead how the arrival of a stranger-guest unsettles and reshapes the relations between those already *at home*. Through *Airbnb*, relationships may expand from being merely familial to becoming entrepreneurial, bringing about a whole new set of responsibilities *towards* each other. Moreover, in some cases, renting out home through *Airbnb* has provoked in some households a re-division of labour in "making-home" *among* each other. Commodifying home has also changed the ways in which people relate to each other and their home. Home, here, produces its identity and continuously constructs its past through shifting social relations. This echoes Massey's (1992, p. 14) critique of the dominant notions of home as an "internally produced, essential past"; a place to which we supposedly "belong". While a place like home may have a character of its own, this is not a unified identity which everyone shares (Massey 1994, p. 169). This raises important questions about how "homes", in the sharing economy of *Airbnb*, are conceived, presented, performed and put into circulation, including the related value production. The platform rhetoric, and the calculative rationalities that underlie its review and rating systems, is supposed to provide a set of objective information concerning the "qualities" of individual homes and of those who perform them locally. Such rather a-historical and a-politicized understanding of home fails to take into consideration that "home" is constituted through the everyday practices of those who inhabit them and, in the case of *Airbnb*, also through each new encounter between hosts and guests, at times leading to contested and difficult relations generated by the very fact of sharing some of the most intimate spaces.

The everyday "becoming" of home, thus, does not lend itself well to sharp categorizations or essentialized qualifications, as the ones provided by *Airbnb*'s review and rating systems. The homes where I stayed in Sofia did not share a singular element of "local" identity; on the contrary, their meaning and the related performances were continuously redefined through the changing social relations and practices that the encounters between *Airbnb* hosts and guests produced on a daily basis. In line with Butler's account of the emancipatory potential of thinking through "becoming", home in this project has emerged less as a "local" object to qualify and rank, and more like a negotiated and sometimes contested process: not a place to return to or a point of arrival, rather a point of departure for both hosts and guests.

References

Accarigi, Ilaria Vanni. 2017. "Transcultural Objects, Transcultural Homes." In *Reimagining Home in the 21st Century*, 192–206. Edward Elgar Publishing. https://doi.org/10.4337/9781786432933.00022.

Adkins, Lisa, and Eeva Jokinen. 2008. "Introduction: Gender, Living and Labour in the Fourth Shift." *NORA—Nordic Journal of Feminist and Gender Research* 16 (3): 138–49. https://doi.org/10.1080/08038740802300947.

Allen-Collinson, Jacquelyn. 2013. "Autoethnography as the Engagement of Self/Other, Self/Culture, Self/Politics, Selves/Futures." In *Handbook of Autoethnography*, edited by Stacy Linn Holman Jones, 281–99. Walnut Creek, CA: Left Coast Press.

Andersson Cederholm, Erika, and Johan Hultman. 2010. "The Value of Intimacy—Negotiating Commercial Relationships in Lifestyle Entrepreneurship." *Scandinavian Journal of Hospitality and Tourism* 10 (1): 16–32. https://doi.org/10.1080/15022250903442096.

Baltova, Stela, and Albena Vutsova. 2021. "Setting the Stage of the Sharing Economy: The Case of Bulgaria." In *The Collaborative Economy in Action: European Perspectives*, edited by Andrzej Klimczuk, Vida Česnuitytė, and Gabriela Avram, 75–89. Limerick: University of Limerick.

Bate, Bronwyn. 2018. "Understanding the Influence Tenure Has on Meanings of Home and Homemaking Practices." *Geography Compass* 12 (1): e12354. https://doi.org/10.1111/gec3.12354.

Baxter, Richard, and Katherine Brickell. 2014. "For Home Un Making." *Home Cultures* 11 (2): 133–43. https://doi.org/10.2752/175174214X13891916944553.

Bialski, Paula. 2012. "Technologies of Hospitality: How Planned Encounters Develop Between Strangers." *Hospitality and Society* 1 (3): 245–60. https://doi.org/10.1386/hosp.1.3.245_1.

Blunt, Alison, and Robyn Dowling. 2006. *Home*. Abingdon: Routledge. https://doi.org/10.4324/9780429327360.

Blunt, Alison, and Varley, Ann. 2004. Geographies of Home. *Cultural Geographies* 11 (1): 3–6. https://doi.org/10.1191/1474474004eu289xx.

Botsman, Rachel, and Roo Rogers. 2010. *What's Mine Is Yours: The Rise of Collaborative Consumption*. New York: Harper Collins.

Bowlby, Sophie, Susan Gregory, and Linda McKie. 1997. "'Doing Home': Patriarchy, Caring, and Space." *Women's Studies International Forum* 20 (3): 343–50. https://doi.org/10.1016/S0277-5395(97)00018-6.

Brickell, Katherine. 2014. "'Plates in a Basket Will Rattle': Marital Dissolution and Home 'Unmaking' in Contemporary Cambodia." *Geoforum* 51 (January): 262–72. https://doi.org/10.1016/j.geoforum.2012.12.005.

Brickell, Katherine, Melissa Fernández Arrigoitia, and Alexander Vasudevan, eds. 2017. *Geographies of Forced Eviction*. London: Palgrave Macmillan. https://doi.org/10.1057/978-1-137-51127-0.

Butler, Judith. 1986. "Sex and Gender in Simone de Beauvoir's Second Sex." *Yale French Studies* (72): 35. https://doi.org/10.2307/2930225.

Butler, Judith. 1988. "Performative Acts and Gender Constitution: An Essay in Phenomenology and Feminist Theory." *Theatre Journal* 40 (4): 519. https://doi.org/10.2307/3207893.

Butler, Judith. 1990. *Gender Trouble*. London and New York: Routledge.

Butler, Judith. 2004. *Undoing Gender*. London and New York: Routledge.

Butler, Judith. 2011. *Bodies That Matter. On the Discursive Limits of Sex*. London and New York: Routledge.

Chesky, Brian. 2014. "Belong Anywhere." Airbnb. http://blog.atairbnb.com/belong-anywhere.

Di Domenico, Marialaura, and Paul A. Lynch. 2007. Host/Guest Encounters in the Commercial Home. *Leisure Studies* 26 (3): 321–38. https://doi.org/10.1080/02614360600898110.

Dowling, Emma. 2012. "The Waitress." *Cultural Studies ↔ Critical Methodologies* 12 (2): 109–17. https://doi.org/10.1177/1532708611435215.

Dupuis, Ann, and David C. Thorns. 1998. "Home, Home Ownership and the Search for Ontological Security." *The Sociological Review* 46 (1): 24–47. https://doi.org/10.1111/1467-954X.00088.

Duyvendak, Jan Willem. 2011. *The Politics of Home: Belonging and Nostalgia in Western Europe and the United States*. Palgrave Macmillan. https://doi.org/10.1057/9780230305076.

Edelman, Benjamin, Michael Luca, and Dan Svirsky. 2017. "Racial Discrimination in the Sharing Economy: Evidence from a Field Experiment." *American*

Economic Journal: Applied Economics 9 (2): 1–22. https://doi.org/10.1257/app.20160213.

Edensor, Tim. 2001. "Performing Tourism, Staging Tourism." *Tourist Studies* 1 (1): 60.

Ellingsen, Winfried Georg, and Knut Hidle. 2013. "Performing Home in Mobility: Second Homes in Norway." *Tourism Geographies* 15 (2): 250–67. https://doi.org/10.1080/14616688.2011.647330.

Fahs, Breanne. 2017. "The Dreaded Body: Disgust and the Production of 'Appropriate' Femininity." *Journal of Gender Studies* 26 (2): 184–96. https://doi.org/10.1080/09589236.2015.1095081.

Gant, Agustín Cócola. 2016. "Holiday Rentals: The New Gentrification Battlefront." *Sociological Research Online* 21 (3): 112–20. https://doi.org/10.5153/sro.4071.

Germann Molz, Jennie. 2007. "Cosmopolitans on the Couch: Mobile Hospitality and the Internet." In *Mobilizing Hospitality: The Ethics of Social Relations in a Mobile World*, edited by Jennie Germann Molz and Sarah Gibson, 65–83. Aldershot: Ashgate.

Germann Molz, Jennie. 2008. "Global Abode." *Space and Culture* 11 (4): 325–42. https://doi.org/10.1177/1206331207308333.

Germann Molz, Jennie. 2013. "Social Networking Technologies and the Moral Economy of Alternative Tourism: The Case of Couchsurfing.Org." *Annals of Tourism Research* 43 (October): 210–30. https://doi.org/10.1016/j.annals.2013.08.001.

Germann Molz, Jennie. 2014. "Toward a Network Hospitality." *First Monday* 19 (3). https://doi.org/10.5210/fm.v19i3.4824.

Gill, Rosalind, and Andy Pratt. 2008. "In the Social Factory?" *Theory, Culture & Society* 25 (7–8): 1–30. https://doi.org/10.1177/0263276408097794.

Goffman, Erving. 1956. *The Presentation of Self in Everyday Life*. Edinburgh: University of Edinburgh.

Gorman-Murray, Andrew. 2006. "Gay and Lesbian Couples at Home: Identity Work in Domestic Space." *Home Cultures* 3 (2): 145–67. https://doi.org/10.2752/174063106778053200.

Gregson, Nicky, and Michelle Lowe. 1995. "'Home'-Making: On the Spatiality of Daily Social Reproduction in Contemporary Middle-Class Britain." *Transactions of the Institute of British Geographers* 20 (2): 224. https://doi.org/10.2307/622433.

Gregson, Nicky, and Rose Gillian. 2000. Taking Butler Elsewhere: Performativities Spatialities and Subjectivities. *Environment and Planning D: Society and Space* 18 (4): 433–52. https://doi.org/10.1068/d232.

Gyimóthy, Szilvia. 2017. "Networked Cultures in the Collaborative Economy BT." In *Collaborative Economy and Tourism: Perspectives, Politics, Policies and Prospects*, edited by Dianne Dredge and Szilvia Gyimóthy, 59–74. Cham:

Springer International Publishing. https://doi.org/10.1007/978-3-319-517
99-5_5.

Haldrup, Michael. 2009. "Banal Tourism? Between Cosmopolitanism and Orien-
talism." In *Cultures of Mass Tourism: Doing the Mediterranean in the Age
of Banal Mobilities*, edited by Pau Obrador Pons, Mike Crang, and Penny
Travlou, 53–74. London and New York: Routledge.

Haldrup, Michael. 2017. "Souvenirs: Magical Objects in Everyday Life."
Emotion, Space and Society 22 (February): 52–60. https://doi.org/10.1016/
j.emospa.2016.12.004.

Haldrup, Michael, and Jonas Larsen. 2006. "Material Cultures of Tourism."
Leisure Studies 25 (3): 275–89. https://doi.org/10.1080/026143606006
61179.

Haraway, Donna. 1991. *Simians, Cyborgs and Women: The Reinvention of
Nature*. London: Free Association Books.

Hardt, Michael. 1999. "Affective Labor." *Boundary 2* 26 (2): 89–100. http://
www.jstor.org/stable/303793.

Hardt, Michael, and Antonio Negri. 2000. *Empire*. Cambridge MA: Harvard
University Press.

Harvey, David. 2008. "The Right to the City." *New Left Review*
53: 23–40. https://newleftreview.org/issues/ii53/articles/david-harvey-the-
right-to-the-city.

Hirt, Sonia. 2006. "Post-Socialist Urban Forms: Notes From Sofia." *Urban
Geography* 27 (5): 464–88. https://doi.org/10.2747/0272-3638.27.5.464.

Hochschild, Arlie Russell. 1997. *The Time Bind: When Home Becomes Work and
Work Becomes Home*. New York: Henry Holt.

Isaksen, Lise Widding. 2002. "Toward a Sociology of (Gendered) Disgust."
Journal of Family Issues 23 (7): 791–811. https://doi.org/10.1177/019251
302236595.

Jarvis, Helen, and Andy C. Pratt. 2006. "Bringing It All Back Home: The Extensification and 'Overflowing' of Work." *Geoforum* 37 (3): 331–39. https://
doi.org/10.1016/j.geoforum.2005.06.002.

Johnston, Lynda, and Robyn Longhurst. 2009. *Space, Place, and Sex: Geographies
of Sexualities*. Boulder CO: Rowman & Littlefield.

Kaneva, Nadia, and Delia Popescu. 2011. "National Identity Lite." *International
Journal of Cultural Studies* 14 (2): 191–207. https://doi.org/10.1177/136
7877910382181.

Karlsson, Logi, Astrid Kemperman, and Sara Dolnicar. 2017. "May I Sleep in
Your Bed? Getting Permission to Book." *Annals of Tourism Research* 62
(January): 1–12. https://doi.org/10.1016/j.annals.2016.10.002.

Kirshenblatt-Gimblett, Barbara. 1998. *Destination Culture: Tourism, Museums,
and Heritage*. Berkeley and Los Angeles: University of California Press.

Knox, Paul, and Steven Pinch. 2014. *Urban Social Geography*. London and New York: Routledge.

Koycheva, Lora. 2016. "When the Radical Is Ordinary: Ridicule, Performance and the Everyday in Bulgaria's Protests of 2013." *Journal of Contemporary European Studies* 24 (2): 240–54. https://doi.org/10.1080/14782804.2016.1170002.

Larsen, Jonas. 2008. "De-exoticizing Tourist Travel: Everyday Life and Sociality on the Move." *Leisure Studies* 27 (1): 21–34. https://doi.org/10.1080/02614360701198030.

Lloyd, Justine, and Ellie Vasta. 2017. *Reimagining Home in the 21st Century*. Cheltenham: Edward Elgar Publishing.

Lynch, P., A. J. McIntosh, and H. Tucker, eds. 2009. *Commercial Homes in Tourism: An International Perspective*. London and New York: Routledge.

MacCannell, Dean. 2013. *The Tourist: A New Theory of the Leisure Class*. Berkeley and Los Angeles.: University of California Press.

Mallett, Shelley. 2004. "Understanding Home: A Critical Review of the Literature." *Sociological Review* 52 (1): 62–89. https://doi.org/10.1111/j.1467-954x.2004.00442.x.

Marston, Sallie. 2004. "A Long Way from Home: Domesticating the Social Production of Scale." In *Scale and Geographic Inquiry*. Wiley Online Books. https://doi.org/10.1002/9780470999141.ch9.

Massey, Doreen. 1991. "A Global Sense of Place." *Marxism Today*, June: 24–29.

Massey, Doreen. 1992. "A Place Called Home?" *Formations* 17: 3–15.

Massey, Doreen. 1994. *Space, Place and Gender*. Minneapolis: University of Minnesota Press.

McDowell, L. 2004. *Gender, Identity and Place: Understanding Feminist Geographies*. Cambridge: Polity Press.

Meah, Angela, and Peter Jackson. 2013. "Crowded Kitchens: The 'Democratisation' of Domesticity?" *Gender, Place & Culture* 20 (5): 578–96. https://doi.org/10.1080/0966369X.2012.701202.

Minca, Claudio, and Tim Oakes, eds. 2012. *Real Tourism: Practice, Care, and Politics in Contemporary Travel Culture*. London and New York: Routledge.

Minchev, Ognyan. 2013. "Russia's Energy Monopoly Topples the Bulgarian Government." German Marshall Fund of the United States.

Mitchell, Katharyne, Sallie A. Marston, and Cindi Katz. 2012. "Life's Work: An Introduction, Review and Critique." In *Life's Work*, 1–26. Chichester: Wiley. https://doi.org/10.1002/9781444397468.ch.

Mohanty, Chandra Talpade. 2003. *Feminism Without Borders*. Duke University Press. https://doi.org/10.1215/9780822384649.

Morgan, Nigel, and Annette Pritchard. 2005. "On Souvenirs and Metonymy: Narratives of Memory, Metaphor and Materiality." *Tourist Studies* 5 (1): 29–53. https://doi.org/10.1177/1468797605062714.

Morrison, Carey-Ann. 2012. "Heterosexuality and Home: Intimacies of Space and Spaces of Touch." *Emotion, Space and Society* 5 (1): 10–18. https://doi.org/10.1016/j.emospa.2010.09.001.

Neychev, Nikolai. 2017. "Airbnb or Long Term Rental." *Kapital* (April 2017). https://www.capital.bg/biznes/moiat_kapital/2017/04/08/2949437_airbnb_ili_dulgosrochen_naem/

Picard, David, and Sonja Buchberger. 2013. *Couchsurfing Cosmopolitanisms.* Edited by David Picard and Sonja Buchberger. transcript Verlag. https://doi.org/10.14361/transcript.9783839422557.

Pratt, Marie Louise. 2008. Imperial Eyes: Travel Writing and Transculturation. London and New York: Routledge.

Richards, Greg. 2017. "Sharing the New Localities of Tourism." In *Collaborative Economy and Tourism: Perspectives, Politics, Policies and Prospects,* 169–84. Cham: Springer International Publishing. https://doi.org/10.1007/978-3-319-51799-5_10.

Roelofsen, Maartje, and Claudio Minca. 2018. "The Superhost. Biopolitics, Home and Community in the Airbnb Dream-World of Global Hospitality." *Geoforum* 91: 170–81. https://doi.org/10.1016/j.geoforum.2018.02.021.

Russo, A. P., and G. Richards. 2016. *Reinventing the Local in Tourism: Producing, Consuming and Negotiating Place.* Bristol: Channel View Publications.

Scarles, Caroline. 2010. "Where Words Fail, Visuals Ignite." *Annals of Tourism Research* 37 (4): 905–26. https://doi.org/10.1016/j.annals.2010.02.001.

Schor, Juliet B. 2017. "Does the Sharing Economy Increase Inequality Within the Eighty Percent?: Findings from a Qualitative Study of Platform Providers." *Cambridge Journal of Regions, Economy and Society* 10 (2): 263–79. https://doi.org/10.1093/cjres/rsw047.

Spry, Tami. 2001. "Performing Autoethnography: An Embodied Methodological Praxis." *Qualitative Inquiry* 7 (6): 706–32. https://doi.org/10.1177/107780040100700605.

Sweeney, Majella, and Paul A. Lynch. 2007. "Explorations of the Host's Relationship with the Commercial Home." *Tourism and Hospitality Research* 7 (2): 100–108. https://doi.org/10.1057/palgrave.thr.6050042.

Tsolova, Tsvetelia. 2013. "Bulgarian Protests for Cheaper Energy Intensify." Reuters World News. http://www.reuters.com/article/us-bulgaria-government-protests/bulgarian-protests-for-cheaper-energy-intensify-idUSBRE91N06D20130224.

Veijola, S., and A. Valtonen. 2007. "The Body in Tourism Industry." In *Tourism and Gender: Embodiment, Sensuality and Experience,* 13–31. Wallingford: CABI. https://doi.org/10.1079/9781845932718.0013.

Veijola, Soile, Jennie Germann Molz, Olli Pyyhtinen, Emily Höckert, and Alexander Grit. 2014. "Introduction: Alternative Tourism Ontologies." In

Disruptive Tourism and Its Untidy Guests, 1–18. London: Palgrave Macmillan. https://doi.org/10.1057/9781137399502_1.

Young, I. M. 1997. *Intersecting Voices: Dilemmas of Gender, Political Philosophy, and Policy*. Princeton, NJ: Princeton University Press.

Zuev, Dennis. 2013. "Couchsurfing Along the Trans-Siberian Railway and Beyond: Cosmopolitan Learning through Hospitality in Siberia." *Sibirica* 12 (1): 56–82. https://doi.org/10.3167/sib.2013.120103.

CHAPTER 5

Datafication of Everyday Life and Bodies

Abstract This chapter investigates the effects of *Airbnb*'s operations and politics at the scale of individual lives and the body. It critically analyses how the concepts of "hospitality" and "life" are (re-)defined through the platform. By examining *Airbnb*'s use of data management and metrics it shows how the platform qualifies and shapes the ordinary lives of its individual hosts. Drawing on a qualitative content analysis of the platform's applications as well as in-depth interviews with *Airbnb* hosts and forum discussions the chapter discusses how people understand and resist how *Airbnb* manages and instrumentalizes the data that is collected on part of their efforts.

Keywords Datafication · Bodies · Biopolitics · *Airbnb* · Digital infrastructure

INTRODUCTION

Tourists' bodies continuously interact with technologies such as cameras, smartphones, portable computers, navigation- and digital devices, significantly influencing how people travel and understand the places they visit (Ash et al. 2018). Travelling and hosting bodies in tourism are also increasingly digitized *through* technology: *digital bodies* arise from vast

© The Author(s), under exclusive license to Springer Nature   99
Switzerland AG 2022
M. Roelofsen, *Hospitality, Home and Life in the Platform Economies
of Tourism*, https://doi.org/10.1007/978-3-031-04010-8_5

amounts of detailed data generated through a range of devices that record people's movements through space, geolocations, behaviour, communication, appearance and many other aspects of their daily lives (Lupton 2016). The individuals generating this information are not, however, the sole beneficiaries of these processes of datafication. As argued in chapter two, personal information has become bestowed with commercial and managerial value serving a digital data knowledge economy (ibid.). While *Airbnb* is not a digital device, its website and mobile applications operate on digital devices that are fed with personal information people knowingly and unknowingly share about themselves. "Data is the lifeblood of the business", according to Riley Newman, former Head of Data Science at *Airbnb* (Rosebush 2014). This slogan provokes thinking on how exactly digitized lives and homes are given value in *Airbnb*'s platform economy. What are the types of data-driven technologies, models and frameworks that underpin the platform and what are the political implications of such data gathering practices? When looking at existing scholarship, critical accounts on the political implications of *Airbnb*'s data harvesting practices have been relatively scarce. Bialski (2016), in analyzing *Airbnb*'s digital infrastructure, argues that unravelling the aesthetic design of the platform allows for certain power structures to become visible. For example, hosts are nudged to advertise their homes in ways that invite a specific "way of looking" by uploading photos to the platform that boast "a certain whitewashed aesthetic of homes" (ibid., p. 47). For Paula Bialski, this competition for authenticity and "coziness" is crucially supported by and entangled in the various digital artefacts and structures that enable the platform to work. Other studies challenge the "logic" of reviewing and rating and analyze the politics enabled through such technologies. Rather than "creating trust" between strangers, the market-based reputation that *Airbnb* users build up is "often about control, manipulation, and discipline rather than transparency and accountability" (O'Regan and Choe 2017, p. 4). In a biopolitical analysis of *Airbnb*, Claudio Minca and I have investigated some of the key technologies and calculative rationalities that drive the making of these digital global communities. We have examined how specific understandings of life and the "spatialities of the home" are central to the quantification/qualification of living spaces generated by the platform (Roelofsen and Minca 2018; Minca and Roelofsen 2021). In the sections below, the main results of these studies are presented.

AIRBNB DATAFICATION PROCESSES

Incorporated in *Airbnb*'s software are a multitude of tools that supply and harvest data on individuals and their everyday spatialities. Together, they continuously create and recreate "digital data assemblages" (Lupton 2016, 2017) of hosting and guest bodies (*bios*) and places (*geos*) alike. *Airbnb* uses such data assemblages to monitor, control and discipline interactions on/through its platform, as well as to modify, privilege or reject specific bodies and spaces through a myriad of practices. Set up as a social network, the *Airbnb* platform contains millions of user profiles. As "data assemblages" (Lupton 2016), these profiles are partial and mutable portraits of *Airbnb* users, configured through the collection of textual, numeric, audio-visual and geolocational data (among many other types of data). Becoming part of *Airbnb*'s networked economy requires prospective users to submit plenty of personal data, including: an email address, first and last name, date of birth, proof of identity, a clear, front-face profile photo, a verified phone number and credit card details. Optionally, users may also indicate where they went to school, the languages they speak and their profession. What follows is a process by which hosts and guests textually and visually detail their profiles through biographies, videos and symbols. *Airbnb* recommends written autobiographical descriptions to include: "things you like", "5 things you can't live without", "favorite travel destinations, books, movies, shows, music, food", "what it's like to have you as a guest or host", "style of traveling/hosting" and a "life motto". *Airbnb*'s former hospitality Toolkits for hosts have proposed even more specific ways to construct and represent one's "authentic" digital Self, in order to draw in "guests with similar tastes and interests" and prevent guests from having disappointing experiences (Airbnb, n.d.). Users can sync their *Airbnb* profiles with profiles on other social networking sites such as Facebook or LinkedIn. The digital verification technology *Jumio*, which is incorporated in the platform, allows for real-time verification of the uploaded forms of identity. After being verified, a statement of the successfully completed verification process is automatically attached to the digital profile of the host or guest and made publicly viewable. Once hosts and guests complete a stay, their profiles automatically become populated by badges and symbols that indicate their accomplishments. These accomplishments are often quantitatively represented through the platform. For example, the number of reviews or recommendations the host or guest has received and, if applicable, the achievement of à Superhost status, which, again relies on the achievement of a certain quantity of values. The profile page ends with a display of all the written, qualitative reviews that the host or guest has received. This display not only "shows and speaks" of what to expect, it also *does* (Kirshenblatt-Gimblett 1998, p. 6), it plays an important role in the production of an "authentic" image of both hosts and guests and how they differentiate themselves among a community of hosts and guests in terms of their performance.

Upon completion, the publicly viewable profile page includes the member's first name, a thumbnail profile picture (only visible to others in high definition when logged in), verified ID information, year of registration, preferred language(s) and place and country of residence. Hosts additionally profile their listing(s) through photos and videos and enter a description that may include what guests may expect in terms of pricing, property/room type, number of bedrooms, beds and baths, cleaning procedures, amenities, location, check-in/out times, cancellation policy, home safety features. They may also post information on co-hosts, fellow inhabitants, the neighbourhood, ways of getting around and the desired level of interaction with guests. The data that *Airbnb* collects about the lives of hosts and guests through profiling alone is therefore vast and of a sensitive and intimate nature. These personal details, along with other data produced by *Airbnb* users, are then stored on cloud computing databases.

SILENT DATAFICATION AND THE WORK OF BEING WATCHED

Other processes of datafication through the *Airbnb* application often silently operate in the background of the users' computers, smartphones and other digital devices. According to *Airbnb*'s privacy policy (2019a), "these activities are carried out based on *Airbnb*'s legitimate interest in ensuring compliance with applicable laws and our Terms, preventing fraud, promoting safety, and improving and ensuring the adequate performance of our services". Users who interact with the application are in fact subjected to digital surveillance: *Airbnb* monitors and records the movement and behaviour of these users, often without their knowledge or awareness. Andrejevic has referred to this as "the work of being watched": a form of unpaid labour by which users willingly or unknowingly "submit themselves to monitoring practices that generate economic value in the form of information commodities" (Andrejevic 2007, p. 304). For example, unless this function is deliberately disabled on the device where the *Airbnb* application is installed and accessed, the platform collects Geolocation Information, tracking hosts' and guests' location through their IP address or the device's GPS (Airbnb 2019a). *Airbnb* also gathers information about the users' interactions, such as the pages and content viewed, searches for listings, and bookings made. Online communication taking place among users, and between users and *Airbnb*, is also recorded, with the message content being scanned, reviewed and analyzed. Data are collected on the hosts' average response time to booking inquiries and new messages sent out to them by guests, and then factored in the ranking of listings in the search results. About average response time, the *Airbnb* company writes:

> [B]ecause making travel plans can be complicated or time-sensitive [...] responses need to be dependable and quick. That's why we calculate response rates and response times and show them on [hosts'] listing pages. We want guests to know what to expect when they reach out to a host. And we ask hosts to stay focused on swift, reliable communication, because we know it will help them confirm more bookings. We want to make sure the meaning of these metrics is clear so that hosts know how to stand out. (Airbnb 2014).

Airbnb's Review and Rating Applications

Other applications incorporated in the platform, such as the review and rating system, gather information on the expression of feelings and on the emotional needs and desires of hosts and guests. These expressions are predominantly the result of interactions among individuals. Over the span of more than 10 years, the review and rating application on *Airbnb* has congregated millions of written testimonies of *Airbnb* experiences all around the world. These reviews and ratings are also the primary source of the users' digital reputation, which in turn has become vastly important for how hosts and guests understand each other, as well as in succeeding in *Airbnb*'s platform economy. Reviews and ratings report on the supposed "quality" of hosts' and on the guests' emotional-, caring- and affective capacities in providing each other with an *Airbnb* experience, as well as on the "quality" of the homes and neighbourhoods where these experiences take place. The insistence on the production of positive feelings is engrained in almost every step of the review process. After each stay, *Airbnb* asks guests to provide the next guests with an account of what they "loved" about their host's place through a written testimony. They are also encouraged to rank their hosts' efforts on a scale of 1 to 5 stars along several parameters, including: Overall Experience, Cleanliness, Accuracy, Communication, Check-in, Value and Location. Additionally, guests are asked to confirm if certain Amenities were present on the property and to rank the home where they stayed in a range starting from "Budget" (limited amenities and minimal furnishing) all the way up to "Upscale" (beautiful space with high-end amenities and decor). Hosts, in a similar vein, are asked to rank their guests' efforts along the parameters of Communication, Cleanliness, Observance of House Rules, and whether they would host the guest again. The majority of these parameters aim to collect information on the behaviour of individual hosts and guests: "How clearly did the guest communicate their plans, questions and concerns? How clean was the guest? How responsive and accessible was the host before and during the stay? How observant was the guest of the house rules?".

Hosts and guests are also expected to behave according to *Airbnb*'s "Community Standards". These standards "help guide behavior and codify the values that underpin" the *Airbnb* community in terms of "trust and safety" (Airbnb 2019b). The five "central pillars" on which the Standards rest include: "Safety", "Security", "Fairness", "Authenticity" and "Reliability", each comprising a subset of categories explaining how hosts and guest are to behave around each other and in each other's homes. For example, on "Authenticity" *Airbnb* provides detailed instructions on how not to misrepresent oneself or the spaces within which the experience takes place. The platform also invites hosts to provide experiences that are "not merely transactions", as "*Airbnb* experiences should be full of delightful moments and surprising adventures" (ibid.). "Reliability" can be achieved by engaging in timely communication, asking hosts to respond to their guests' messages within 24 h. The instructions on how to behave are at times (deliberately?) vague, leaving room for reinter-pretation and, at the same time, expansion of the arena of disciplinary behaviour.

Although reviews and ratings are voluntarily submitted to the platform, *Airbnb* continuously encourages its users via multiple emails and text messages to upload "thorough reviews" in a timely manner in order to "aid the decision-making of future guests and hosts" (Airbnb 2019c). For a host, not being reviewed by a guest after their stay, may result in not being promoted to the "Superhost" status—a status that indicates a 4.8 or higher average overall rating based on reviews from at least 50% of their guests in the past year. Successful Superhosts attract more guests by being "featured to guests" in search results, in this way increasing their chances of being booked (Airbnb 2019d). When Superhosts fail to get reviewed according to the 50% benchmark in new assessment rounds, they risk having their Superhost status revoked (Airbnb 2019e, 2019f). This may lead to less bookings and a lower earning potential. In other words, not to report or not to be reported on is considered adverse behaviour in the *Airbnb* economy and is directly and indirectly sanctioned.

Today's feedback- and assessment mechanisms underpinning countless digital (tourism) platforms may have significant effects on how individuals think of themselves—as ranked "quantified Selves" (Lupton 2016). On the one hand, keeping track of personal digital data that underpin people's "scores" may be a strategic means of self-promotion. On the other, data are also algorithmically manipulated and represented in certain forms—shaping behaviour towards specific ends (Pasquale 2015, p. 38). As such, data can also be misused and instrumentalized to target people based on their vulnerabilities (Pasquale 2017).

THE DEEPER POLITICS OF REVIEWS, RATINGS AND RANKINGS

David Beer in *Metric Power* (2016) explores the deeper logics of big data and shows how the presence of metrics in our everyday lives has tremendously intensified in the last 15 years. A growing amount of data is collected on *what* and *how* we do, and the metrics that capture this information increasingly play a role in our lives (Beer 2016). Lupton (2016, p. 2) describes self-tracking as a practice in which people regularly monitor, record and measure elements of their behaviour and/or their bodily functions. As noted above, *Airbnb* hosts and guests are incentivized in various ways to produce data feeding into the metrics on which *Airbnb* thrives. However, the incentive to rate and review goes beyond building reciprocity and trust. Although reviewing and rating is neither obligatory nor compensated, to be reviewed and rated positively is vastly important, to some extent *essential* to stay "alive" in this platform economy. A trustworthy digital Self does not merely rest on documented forms of identification and auto-biography; it emerges through testimonies of both hosting- and guesting performances.

As a form of social regulation, the growing importance of each individual's "digital reputation" is a powerful incentive for the platform's members to act in the "desired" manner, possibly without *Airbnb*'s direct intervention. Such self-discipline is further enforced through the *Airbnb* "dashboard" incorporated in the host's profile page, which provides key metrics of their various expected performance. The dashboard includes an application called "Progress", which monitors, records, organizes, measures, analyses and presents a variety of data on the hosts' behaviour, who are in this way constantly reminded of their "rated" performances of hospitality over time (Airbnb 2019g). "Progress" enables the practice of "self-tracking", allowing users to reflect on certain patterns in their behaviour and accordingly improve their relationships with their guests. Unlike *Airbnb* hosts, *Airbnb* guests are not provided with a tool to self-track their performances, although they are exposed to written reviews after their stay. The guest's metrics, however, are only shared with a specific group of hosts who allow for their listings to be "instantly booked" without prior communication with the respective guest. If hosts "ever rate a guest at 3 stars or below", this guest will not be able to instantly book with the same hosts again (Airbnb 2017). Both hosts and guests, then, are offered (and incentivized to use) digital technologies that render them and others into "quantified selves" along a set of given parameters and, accordingly, allow them to assess whether or not they wish to engage with each other.

The hosts' "dashboard" and "statistics tool" are examples of built-in applications that, besides inspiring a culture of self-tracking, imply the necessity to perform introspection and self-improvement. By showing less-than-perfect reviews/ratings hosts are encouraged to clean their homes more thoroughly, to be clearer and faster in communicating, to be more accurate and truthful and to accept more reservations. The platform's expectation that hosts should be working towards a Super Self are underpinned by a "notion of incompleteness and a set of moral obligations concerning [...] contemporary ideas about selfhood and citizenship" (Lupton 2016, p. 68). In the case of *Airbnb*, such ideas are contextualized with reference to the sharing economy, implying a kind of global citizenship concerned with sharing virtually everything that one embodies or owns, in ways that supposedly "enable greater efficiency and access" (for an enthusiastic endorsement of this logic see Botsman 2015).

DISCIPLINARY AND REGULATORY POWER

Taking on biopolitical lenses, the above described standards and measurements somewhat reflect what Foucault (2003, p. 249) has famously introduced as two distinct forms of power: a disciplinary power, aimed at the individual (host and/or guest); and a regulatory power, aimed at the level of the community (of *Airbnb* citizens). While the first form of power should increase the (economic) productivity of the individual, the second is concerned with preventing the community from risk and protecting it from potential danger. Both forms of power adopt their own instruments and operate with different objectives but together they form interrelated components of a "political technology" that controls as much the individual as the workings of the community. In establishing "quality", the *Airbnb* community facilitates self-discipline and self-regulation. Hosts and guests are continuously encouraged through emails and notifications to provide opinions on their experiences, and improve their performances through self-tracking. According to the promotional material, *Airbnb* operates review and rating systems to assist "the community [of hosts and guests] to make better decisions" (*Airbnb*). Daily interactions among users of the platforms are monitored by Trust and Safety Teams that offer customer support—and are a self-proclaimed "community's vanguard" (ibid.). The trustworthiness of hosts and guests is presented in the form of an aggregate of their assessed behaviour and of their social relations with other members of the *Airbnb* "community". At the same time, it operates as a machinery collecting information on individuals and their places, in order to quantify their qualities and incorporate them into a true (bio-geo)metrics. In relation to travel, thus reviews and ratings generate actions and reactions and have effectively become co-constitutive of the multiple identities of tourist destinations and of new kinds of place-making (see Baka 2015, p. 151). The struggle over which tourist places are made/unmade and booked/shun increasingly thus depends on the ordering devices that "weigh" their quality via reviews and produce rankings. Such subjective opinions are rendered objective by mathematical algorithms used by these platforms, which convert into calculative rationalities and related metrics the reputational economies of its affiliates.

Like other networks in the platform economy, *Airbnb* thrives within the "reputational turn". In the process of reciprocal ranking among hosts and guests, qualitative distinctions are translated into quantitative ones, a process that "actively works to depersonalize and de-particularize the very activities being measured" (Hearn 2010, p. 428). This, however, does not prevent *Airbnb* hosts and guests from uploading reviews and ratings not conforming to the level of "objectivity" or "sincerity" that the company aspires to obtain. While *Airbnb* promises that reviews and ratings provide some "transparency" or "truth" about individual members and the spaces where the *Airbnb* experiences take place (Airbnb 2019h), the reviewers' sentiment about those same experiences are not unaffected "by already existing class, gender, race and other social relations" (Hearn 2010, p. 433). Providing feedback is not "only ever motivated by an honest desire to do good" (ibid.), since reviews and ratings are oftentimes the result of diverse, biased or even conflicting understanding of "quality" among hosts and guests. As noted in a recent post by an *Airbnb* administrator:

'Reviews are so important. They not only impact the success of your business, they're also really personal. We know you put a lot of thought and care into your hospitality and that it's frustrating when you receive a review that is uncharacteristically low—be it a mistake, a misunderstanding, or an unfair assessment. [...] We've invested and will continue to invest a lot of thought and effort into how we can make the review system more fair' (ACC 2018).

Airbnb is thus committed to continue improving its review system and to adopt tools capable of detecting "outlier reviews"—that is, reviews and ratings that do not accurately represent "truthful" feedback (ibid.). What also transpires from the above statement is that reviews and ratings are presented as a taken for granted aspect in the "self-governance" of the platform. Truthful and objective *reporting* on each other's behaviour is considered of fundamental importance in keeping *Airbnb*'s feeling-intermediary credible and "risk-free"—behaviour that the platform is willing to monitor and "correct" when necessary.

Safety, Security and Coercion of Social Behaviour

By using predictive analytics and "machine learning", every *Airbnb* reservation is "scored" for risk before being confirmed (Hakim and Keys 2014), "instantly evaluating hundreds of signals that help [*Airbnb*] flag and investigate suspicious activity before it happens" (Airbnb 2019i). In order to facilitate this, the platform requires data. In a recent interview with Bloomberg, Nick Shapiro, former CIA's Deputy Chief of Staff and White House counterterrorism and homeland security aide to President Obama and former Global Head of Trust and Risk Management[1] at *Airbnb*, has explained that:

> People need to know that they are going to be safe, they have to feel safe. So, we do a number of things, to use technology to do that. We risk-assess each and every reservation. We run global watchlist checks against all of our users worldwide. We background check hosts and guests in the US. But just being safe isn't enough. We use these technologies to also build connections. There's detailed profiles. There's the messaging system where you can learn more about each other. And there's reviews where you can look at previous history. And on top of that, people need to know that they are not alone. *Airbnb* is there for them. (Bloomberg Technology 2017)

What becomes clear from the above interview is that a multitude of data on intimate elements of the members' everyday life is incorporated by *Airbnb* into predictive analytics aiming at anticipating hosts' and guests' potential behaviour. Such behavioural data are of vast importance to enable *Airbnb* in nudging and coercing social behaviour on a large scale towards the most profitable outcome. Importantly, these reviews and ratings rely entirely on the free labour of hosts and guests, whose affective participation is entirely voluntary and unpaid (see Terranova 2000). While often promoted as being fundamental to "attracting new business", the value created through reviews and ratings crucially contributes to the deeper logic of the platform. Data generated by reviews and ratings in fact feed into algorithms that sift and sort them to optimize future transactions, with the main beneficiary being *Airbnb* that draws on this massive data to improve its market appeal and nudge the members' social behaviour towards the company targets. These algorithms also measure how often hosts cancelled a reservation, the amount of stays they hosted

---

[1] Shapiro has moved on to become the Global Head of Crisis Management at *Airbnb*.

and their overall ratings by guests. As such, these categories depend on the hosts' dedication and aptness, to inspire a culture of constant "self-improvement" and to invite them to be virtually on-demand every minute of the day.

The review and rating system must thus be continuously alimented by "embodied data" concerning the behaviour of individuals in order to effectively shape future interactions on the platform, while hosts and guests alike are strongly encouraged to provide such information. These "embodied data" in fact determine how guests, hosts and their homes are ranked in the platform's search index and booking tool, software associated to algorithms that identify, classify, structure and prioritize certain people and certain homes over others. As such, these algorithms are far from being neutral (Kitchin 2017), since they *do* assess homes and individuals according to specific parameters and "values" set by *Airbnb*. As Safiya Noble illustrates in her recent book on Google (2018, p. 2), "some of the very people who are developing search algorithms and architecture are willing to promote sexist and racist attitudes openly at work and beyond, while we are supposed to believe that these same employees are developing 'neutral' or 'objective' decision-making tools".

AIRBNB SEARCH REVEALED: OR, THE IMPORTANCE OF RANKING WELL

On October 21st, 2017, millions of hosts received the monthly *Airbnb* newsletter in their inboxes, which included a link to an online post written by Lizzie, *Airbnb*'s "Online Community Manager". Lizzie wished to provide more information to *Airbnb* hosts about the platform's search algorithm, because, according to the stats, one of the most popular topics on the *Airbnb*-Community fora was: "how *Airbnb* Search works" (ACC 2017). Responding to a large cohort of hosts speculating about how the search algorithm classified and ranked their homes, Lizzie's post revealed some of the underlying ideas driving this powerful calculative device. Lizzie explained that the algorithms respond to a specific set of preferences (e.g. dates, location, etc.) that guests enter into the online booking tool. Based on these preferences, the algorithms sift through existing data on all *Airbnb* homes and on hosts' past performances to rank those that best "match" the guest's criteria.

While Lizzie indeed provided some hints, these were never *too specific*:

'We have an algorithm that looks at over 100 signals to decide how to order listings in search results. Most of those signals have to do with things that guests care about, like positive reviews and great photos. If you think guests might care about it, it probably factors into your ranking!' (ACC 2017).

Similar to the *immunitary* and *control* strategies observed in van Doorn's study (2017) of on-demand platforms such as Uber, Lizzie's comment reveals the strategies employed by *Airbnb*. The first strategy is Lizzie's appeal to the algorithm as an "independent" assessor of *Airbnb* hosts' performances. Designating the algorithm as an objective measure "shields" the company (*Airbnb*) from dealing directly with those who perform the labour of hosting on the platform. In doing so, the platform rids itself from as much liability as possible (van Doorn 2017). Secondly, the post suggests that what guests care about is decisive in the ranking of listings in the *Airbnb* search results. In other words, *Airbnb* outsources quality control to the hosts' "customers"—namely the *Airbnb* guests.

Lizzie then claimed that hosts do not need a perfect listing or an unbeatable location for ranking well and suggested *how* hosts could possibly improve their position in terms of search results. Lizzie insisted in particular on the importance of activating the Instant Booking feature:

> [T]ravelers prefer to use Instant Book because they can book quickly, skip the wait time for hosts to respond, and avoid possibly being rejected. Because of the high booking success for hosts and guests, Instant Book gives your listing a boost in searches. (ibid.)

The main incentive for hosts to use Instant Book derives from not having to franticly maintain what the administrator described as: "welcoming correspondence and strong response metrics". Automated messages would do the work of "responding to the guest in real time", making for the highest possible response rate and a higher rank in search results. A second incentive is the possibility, offered only by Instant Book, to see how other hosts have rated their prospective guests—offering "more peace of mind" (Airbnb 2019j). What Lizzie forgot to mention was the considerable advantages for *Airbnb* in having hosts accept all booking requests without prior consultation. Besides instantly receiving commissions on the booking payment, *Airbnb* conveniently avoids any difficulty that may arise from human interactions between hosts and guests. Instant

Book makes all preliminary and potentially "unruly" social interactions redundant. An algorithm will do the job!

What matters for the sake of the argument is that the responses to Lizzie's post show how many hosts have learned to incorporate but also resist the overall logic of the platform, its capitalization on their home and its power in attributing value to people and their practices.

Janine, a long-time host, for example, clearly expressed her discontent on the ACC. Janine lives with her guests in her home and, because of that, it is vital for her to have the "opportunity to choose" who she is sharing her intimate everyday spaces with. Janine declared that she "does not think Instant Book 'should' be a search factor at all". In fact, her status as a Superhost for over 2 years "should make a difference!" In another response, Sally contended that she does not use Instant Book either, because her place "is not a hotel; it is a home". According to Sally, sometimes her and the guests are not "mutually compatible". Sally noticed that many people do not read her House Rules on her listing page, "so a few e-mails back and forth help create a mutually positive experience". Arguably, the concerns manifested by Janine, Sally and others were not about the fact that people *are* ranked but rather about *how* they were ranked. These concerns clearly emerged from the analysis of the *Airbnb* Search post (ACC 2017) and other materials made available by the platform but also from the interviews. Many hosts in fact complained about the ranking system, since they would have liked to have *more reviews* and *better rankings*, and more personal information on their potential guests.

"I think maybe the system could improve reviewing and rating a little bit. They could require more detailed reviews and references. Because now you have a specific set of categories, like cleanliness and tidiness, etc. And then you read a review based on 200 characters and this is it. And then you want to say something personal to your guests. I understand that often you don't have time to write a proper review and describe in detail your experience with your guests or hosts or whatever. But this is essential especially when you rent a room, the flat is ok but when you rent a room you co-live with this person!" (interview with host Pino 2015).

Some members felt that the guests should be rated as well. Supported by hundreds of "likes", one vocal host replied to a related Administrator-post addressing the "one-off bad reviews" and practices of blackmailing on the platform. According to this host, *Airbnb* guests "will be very critical of a property only because they want a big discount and they will threaten their host with a bad review". One of her *Airbnb* guests recently informed her that the taps in her house were "old fashioned", and that she should have modern ones. The guest also "photographed dust and magnified the pictures and told [the host] that she wanted a refund which [the host] gave to her as [she] did not want a bad rating".

All in all, the analysis revealed the implicit power of *Airbnb*'s algorithms in regulating a specific set of social relationships, even when the members expressed dissatisfaction and concern about their workings. A number of comments—both on the ACC and in person—also revealed how the availability of personal information on the platform may offer the ground for discriminatory or racist behaviour of some members, despite the attempts on the part of *Airbnb* to sanction these practices. In an interview, Emma, a host in her early 30s, reflected on her own routines in selecting guests:

> We try to make a selection [before we accept bookings]. We look through guests' profiles and read the comments of previous hosts. We never host a person without comments or references. I decline a lot of requests just because I don't like to get this feeling that these people only come here for the bed and nothing else. So, we are searching for interesting people. Interested to share and exchange something and spent time together.

Many hosts also admitted having implemented various forms of self-disciplining in order to get a better ranking or at least become visible via the search tool. Proper feedback was considered by many as paramount to become a proper "citizen" of the *Airbnb* global community, something clearly illustrated by host Madeleine during an interview:

> 'Yes, I really want to review all of them because I think it is really important to share what my impressions were about them. Also, to give them kind of feedback for themselves.
>
> Feedback is really important to know where you are and what are you doing well or not so well'.

Another interviewee, Dave, openly declared having changed his behaviour at home to comply with the platform's expected standards and get good reviews. When guests come over, he tries 'to be more calm, quiet', and cleans up after himself using the toilet and common spaces.

> What changes [when guests come over] is that I try to be more calm, quiet. Not to argue a lot with my mother. I would not watch TV louder.

I wouldn't have parties. Or, I'll ask if the person is ok with it. And you should clean all the time after you use something. If you go to the toilet or if you use the kitchen and stuff like that.

Mercedes, not only made a few changes in the home to accommodate potential guests, but also mobilized personal social networks in order to obtain some recommendations, eagerly searching for something that would get her on *Airbnb*'s map of global hospitality:

I did some renovations in my apartment and I bought many things like sheets, you know. You have to prepare a lot of stuff for Airbnb. I was wondering is it going to work or no? You never know. *You need reviews to get guests and need to have guests to have reviews.* And I asked some friends to write me, not reviews, but a recommendation. And step by step I had some guests.

Despite the criticism expressed by many members, the review system is commonly considered reliable and truthful among the interviewees, a fundamental tool to build an individual capital of trust that will be reflected in the ranking. Again, Mercedes noted:

Yeah. I think it is very important. I mean for regular travelers... for example, I wouldn't stay at some place [when] they don't have a review. This is the way to know about some things that the host doesn't mention, like there is no hot water from time to time. So, I think the reviews are very very important.

A third group of responses shows clear awareness of the possibility that some members may use the ranking system at their own advantage or provide "false" or unreliable information. For host Ada,

It's definitely good to have the review system. But as I confessed, I rarely write what I really think. Especially when it's negative. When it's very negative, then I write it. Because when it gets to a point of danger or something like this. I think this is already more security than somewhere else.

Another host argued for a more detailed history of how guests reviewed previous hosts:

I would also suggest that Airbnb take account of the number of stars the guests have given other hosts in the past. Some guests just don't give 5 star reviews no matter what. That should be weighted in the algorithm for host reviews and a correction should apply for this.

Overall, the idea that *Airbnb* has not only impacted the tourist and rental market of many cities, but also the social behaviour of many hosts in some of their most intimate spaces, was supported by clear evidence emerging from our interviews and most of the materials consulted.

Concluding Thoughts: *Airbnb*'s Biopolitics

Airbnb's digital infrastructure is data-driven; it operates by capturing, storing, repurposing and redistributing information about elements of life, home, care, coziness, local culture—all incorporated by the platform as "values"—which are converted into quantitative measures producing a specific kind of hierarchy. Biopolitical technologies such as these not only "contain" elements of life, but also endlessly "qualify" life (Minca 2015). They produce a mapping made of specific representations of life. In the *Airbnb* world, users of the platform are actually mapped out: as individuals, as families, as "home", but also as travellers, as guests, as providers of care and "hospitality". The *Airbnb* ranking of people and homes via its algorithms in fact comprises the incorporation of elements of real life, real homes, real relationships, into the topographical logic of the platform. The qualified quantifications of these real-life elements tend to shape what hospitality is for *Airbnb* and the members of its community, and how these latter should perform in order to match that very calculative rationality—that is, "how to behave" to be highly ranked in a world-made-of-hosts-and-guests.

Yet, the *Airbnb* algorithms are machineries that spin around an empty core. Despite the fact that the ideas of community, home, hospitality, local culture, etc. feeding into the platform's algorithms are linked to real-world contexts—the homes offered are real as are their locations—when they are translated into the *Airbnb* ranking, they tend to become something else, possibly self-referential metaphors based on *Airbnb*'s calculative rationalities. This does not mean that they do not operate as social regulators, quite the contrary. The *Airbnb* algorithms—as shown by this chapter—are in fact part of a biopolitical technology that squeezes value out of a myriad of aspects of everyday life. These aspects are often voluntarily offered by the participants, who willingly put on display a series of intimate and personal elements of their respective lives (and homes) to have them incorporated in the grand metrics of the platform. Many of these hosts actually enjoy being involved in these encounters with their guests, and often interpret their role in ways that resists and somewhat "twists" the rationale of the platform, as shown in interviews discussed above. However, despite these subjective (and sometimes even subversive) interpretations of the interplay between guests and hosts, by feeding "home" and "everyday life" into its metrics the calculative rational of the platform somehow tends to empty them out, to convert them into elements of a world generated by datafication and algorithmic management. And it is precisely for this reason that, *Airbnb*, like all biopolitical machinery, needs endless injections of new life (new homes, new intimate encounters, new hosts and guests), new "stuff" to be put into circulation, new spaces to be incorporated into its broader regulatory system. Indeed, the embodied data collected and actively created by *Airbnb* on part of millions of hosts and guests are an important form of capital, without which *Airbnb* would not be able to operate nor to generate value (see Sadowski 2019).

What is more, and this is a third theoretical proposition, the *Airbnb* logic cannot be taken to its most extreme consequences. This is confirmed by the ideal citizen of the *Airbnb* community, the Superhost, a host ranked 4.8 to 5 stars, who represents a distilled and embodied abstraction of the deeper logic of its algorithms (see Roelofsen and Minca 2018). While each member of this community of hospitable residents should aspire to obtain the condition/status of Superhost, at the same time, this is an endlessly mobile condition, since the rules to maintain it are constantly changed by the platform. The Superhost in fact must remain the horizon towards which all hosts move, but that nobody can actually permanently inhabit. As we have learned from the history of all biopolitical regimes, the principle of "endless improvement" does not produce a perfect(ed) society, because the workings of biopolitics is based on movement, on ever-changing thresholds of inclusion and exclusion (see, Agamben 1998). Viewed in this way, the *Airbnb* platform acts as a biopolitical machinery spinning around an empty core; its objective, in the end, is to reproduce its capitalist economy. There is thus no goal, no point of arrival, no community to be realized, no perfect guest or host.

REFERENCES

Agamben, Giorgio. 1998. *Homo Sacer*. Stanford CA.: Stanford University Press.

Airbnb. n.d. "All Toolkits. Resources for Hosting."

Airbnb. 2014. "Dependable Communication Builds Community."

Airbnb. 2017. "See Your Guest's Star Rating."

Airbnb. 2019a. "Updating Our Terms of Service, Payments Terms of Service and Privacy Policy." https://news.airbnb.com/updating-our-terms-of-service-payments-terms-of-service-and-privacy-policy-2/.

Airbn. 2019b. "Your Safety Is Our Priority." https://www.airbnb.com.au/trust/standards.

Airbnb. 2019c. "How Airbnb Works."

Airbnb. 2019d. "Superhost: Recognizing the Best in Hospitality."

Airbnb. 2019e. "Can I Lose My Superhost Status." https://www.airbnb.com.au/help/article/832/can-i-lose-my-superhost-status.

Airbnb. 2019f. "How Do I Become a Superhost."

Airbnb. 2019g. "How Do I Track My Superhost Status."

Airbnb. 2019h. "How Does Airbnb Help Build Trust between Hosts and Guests."

Airbnb. 2019i. "What Does It Mean When Someone's ID Has Been Checked." https://www.airbnb.com.au/help/article/2356/what-does-it-mean-when-someones-%0Aid-has-been-checked.

Airbnb. 2019j. "Business Is Better with Instant Book." https://www.airbnb.com.au/host/instant.

Andrejevic, Mark. 2007. Surveillance in the Digital Enclosure. *The Communication Review* 10 (4): 295–317. https://doi.org/10.1080/107144207017 15365.

Ash, James, Rob Kitchin, and Agnieszka Leszczynski. 2018. Digital Turn, Digital Geographies? *Progress in Human Geography* 42 (1): 25–43. https://doi.org/10.1177/0309132516664800.

Baka, Vasiliki. (2015). Understanding Valuing Devices in Tourism through "Place-making". *Valuation Studies* 3 (2): 149–180. https://doi.org/10.3384/VS.2001-5992.1532149.

Beer, David. 2016. *Metric Power*. London: Palgrave Macmillan UK. https://doi.org/10.1057/978-1-137-55649-3.

Bialski, Paula. 2016. "Authority and Authorship: Uncovering the Sociotechnical Regimes of Peer-to-Peer Tourism." In *Reinventing the Local in Tourism: Producing, Consuming and Negotiating Place*, edited by Antonio Paolo Russo and Greg Richards, 35–49. Bristol, Blue Ridge Summit: Channel View Publications. https://doi.org/10.21832/9781845415709-005.

Bloomberg Technology. 2017. "How Airbnb Is Using Tech to Build Trust."

Botsman, Rachel. 2015. "Defining The Sharing Economy: What Is Collaborative Consumption—And What Isn't?" *Fastcompany* 27 (May):

1–7. http://www.fastcoexist.com/3046119/defining-the-sharing-economy-what-is-collaborative-consumption-and-what-isnt.

Center, Airbnb Community. 2017. "How Search Works." https://community.withairbnb.com/t5/Hosting/Yourtop-%0Aquestions-about-Airbnb-Search/td-p/509644.

Center, Airbnb Community. 2018. "Airbnb Answers: Protecting You from One-off Bad Reviews." https://community.withairbnb.com/t5/Airbnb-Updates/%0AAirbnb-Answers-Protecting-you-from-one-off-bad-reviews/td-p/822623.

van Doorn, Niels. 2017. Platform Labor: On the Gendered and Racialized Exploitation of Low-Income Service Work in the 'on-Demand' Economy. *Information, Communication & Society* 20 (6): 898–914. https://doi.org/10.1080/1369118X.2017.1294194.

Foucault, M. 2003. *Society Must be Defended: Lectures at the College De France,* 1975–76 (David Macey, trans.). Picador, New York.

Hakim, Naseem, and Aaron Keys. 2014. "Architecting a Machine Learning System for Risk." Medium.

Hearn, Alison. 2010. "Structuring Feeling: Web 2.0, Online Ranking and Rating, and the Digital'reputation'economy." *Ephemera: Theory & Politics in Organization* 10: 421–38. http://www.ephemerajournal.org/contribution/structuring-feeling-web-20-online-ranking-and-rating-and-digital-'reputation'-economy.

Kirshenblatt-Gimblett, Barbara. 1998. *Destination Culture: Tourism, Museums, and Heritage.* Berkeley and Los Angeles: University of California Press.

Kitchin, Rob. 2017. Thinking Critically about and Researching Algorithms. *Information Communication and Society* 20 (1): 14–29. https://doi.org/10.1080/1369118X.2016.1154087.

Lupton, Deborah. 2017. "Digital Bodies." In *Routledge Handbook of Physical Cultural Studies*, edited by Michael L. Silk, David L. Andrews, and Holly Thorpe, 200–208. Abingdon, Oxon; New York, NY: Routledge, 2017. | Series: Routledge. https://doi.org/10.4324/9781315745664.

Lupton, Deborah. 2016. *The Quantified Self*. Cambridge, UK: Polity.

Minca, Claudio. 2015. "The Biopolitical Imperative." In *The Wiley Blackwell Companion to Political Geography*, 165–86. Chichester, UK: John Wiley & Sons, Ltd. https://doi.org/10.1002/9781118725771.ch14.

Minca, Claudio. 2012. "No Country for Old Men." In *Real Tourism: Practice, Care, and Politics in Contemporary Travel Culture*, edited by Claudio Minca and Tim Oakes, 12–37. Routledge. https://doi.org/10.4324/9780203180969-8.

Minca, Claudio, and Maartje Roelofsen. 2021. Becoming Airbnbeings: On Datafication and the Quantified Self in Tourism. *Tourism Geographies* 23 (4): 743–764. https://doi.org/10.1080/14616688.2019.1686767.

Noble, Safiya Umoja. 2018. *Algorithms of Oppression*. New York University Press.

O' Regan, Michael, and Jaeyeon Choe. 2017. "Airbnb and Cultural Capitalism: Enclosure and Control within the Sharing Economy." *Anatolia* 28 (2): 163–72.https://doi.org/10.1080/13032917.2017.1283634

Pasquale, Frank. 2017. "Written Testimony of Frank Pasquale, Professor of Law University of Maryland Before the United States House of Representatives Committee on Energy and Commerce Subcommittee on Digital Commerce and Consumer Protection "Algorithms: How Companies' Decisions," 1–24.

Pasquale, Frank. 2015. "The Algorithmic Self." *The Hedgehog Review* 17 (1): 1–7. http://www.iasc-culture.org/THR/THR_article_2015_Spring_Pasquale.php.

Roelofsen, Maartje, and Claudio Minca. 2018. The Superhost. Biopolitics, Home and Community in the Airbnb Dream-World of Global Hospitality. *Geoforum* 91: 170–181. https://doi.org/10.1016/j.geoforum.2018.02.021.

Rosebush, Steve. 2014. "Airbnb Says Data Is 'Lifeblood' of Fast-Growing Business." *Wall Street Journal*, 2014. https://www.wsj.com/articles/BL-CIOB-4115.

Sadowski, Jathan. 2019. When Data Is Capital: Datafication, Accumulation, and Extraction. *Big Data and Society* 6 (1): 1–12. https://doi.org/10.1177/2053951718820549.

Terranova, Tiziana. 2000. "Free Labor: Producing Culture for the Digital Economy." *Social Text* 18 (2): 33–58. https://muse.jhu.edu/article/31873.

CHAPTER 6

Conclusion

Abstract In this concluding chapter the main findings of the book are summarized. The chapter also provides suggestions for future research on tourism and digital platforms.

Keywords Platforms · Tourism · Hospitality · Digitalization · Data

This book has provided insights into the emergence of digital tourism platforms and how they shape cities, home, hospitality, and everyday life. Empirically, it has focused on a set of digitally enabled geographies that are largely but not exclusively related to the *Airbnb* economy in Sofia, Bulgaria. Through a historical perspective on the digitalization of tourism, the book has argued that while tourism platforms are often labelled "revolutionary" they have necessarily been built upon decades of technological development in computing. These technological developments have assisted the hospitality and tourism sector to grow exponentially since the 1950s. There are obvious differences between the binary computing architectures of the earlier Computer Reservation Systems and today's booking "platforms" but the principles and ideas that have driven their inventions are similar. Relying on methods of data collection and data management, most of them have been developed to sustain market networks at a global scale, with the purpose of increasing

© The Author(s), under exclusive license to Springer Nature
Switzerland AG 2022
M. Roelofsen, *Hospitality, Home and Life in the Platform Economies of Tourism*, https://doi.org/10.1007/978-3-031-04010-8_6

sales and generating profits. They have vigorously re-shaped the distribution and use of information, work processes and tourism- and hospitality labour, global mobility and the concurrent consumption and production of tourism places and cultures.

By means of illustration, I have shown that the digital architecture of *Airbnb* today consists of numerous applications that have long underpinned the most elementary booking- and inventory systems of hotel chains and other accommodation providers. Yet, while they rely on similar means of data collection and management there are major differences between traditional accommodation providers and platforms like *Airbnb*. Platform enterprises do not own any of the rooms and homes that they enlist, nor do they pay for any of the labour that is needed to provide clean, safe, and hospitable space and tourism experiences. Instead, they bring numerous users together in two-sided or multi-sided marketplaces, where they charge commissions for completed transactions and, importantly, capture value from the data that users generate. These multiple users are diverse in terms of their occupation and intent, and they are oftentimes, but not always, new to the demands of the tourism and hospitality profession. They include homeowners, tenants, landlords, investors, domestic workers, neighbours and neighbourhood residents, family members, students, "digital nomads" and, not in the least, tourists, among many others.

Platform enterprises often design and operate their platforms as value-generating networking systems that target spaces and resources which stand "idle" or are assumed to have "excess capacity". Oftentimes, these platforms also draw on existing forms of labour that are, generally speaking, portrayed as "unproductive". Short-term rental platforms, for example, incentivize people to enlist and monetize domestic space as well as forms of housework, care, and emotional labour to accommodate guests. According to one of *Airbnb*'s most recent promotional campaigns, anyone "can host anything anywhere" alluding to the infinite possibilities that people around the globe supposedly have to commercialize everyday life in intimate and residential space (Airbnb 2022). Based on such rhetoric, it would be easy to conclude that enterprises like *Airbnb* merely re-shape cities and everyday life by neoliberal capitalist interest; their motive is to generate individual gain and extract profit. While there is truth in such argument, the book has also shown that these platforms do not operate in a top-down fashion, nor do they operate in a socio-spatial vacuum.

The everyday spaces and spatialities in which tourism platforms become embedded, always already were "made" socially, emotionally, and relationally. Moreover, the assumption that housework and the care that are needed to maintain everyday life at home are "unproductive" just because they were unpaid before the advent of a platform economy is a fundamentally flawed one. Despite being ideologically devalued; housework and care have *always already been* productive and integral to paid work under capitalism (Federici 1975). This is a notion and critique that has long prevailed in feminist literature and an argument that is woven throughout this book. Similarly, "home" as a dwelling, shelter, and site of social and emotional relations, has always been essential to ensure that those whose work *is* remunerated are able to recover and rest, allowing them to physically and emotionally be able to continue working (Blunt and Dowling 2006).

Paradoxically, though, it has become increasingly clear throughout the book that *Airbnb* has "made-home" in many neighbourhoods and cities *at the expense of* already existing homes and livelihoods (Goyette 2021; Roelofsen 2021). As Chapter 2 has shown, the platform has exacerbated urban housing crises around the world by converting long-term housing into short-term rentals, further aggravating the availability and affordability of housing for residents. In other ways, the platform has contributed to feelings of unhomeliness among those residents who now, unwillingly, have to share their livelihoods with short-term visitors and tourists who usually contribute little to the long-lasting relationships and practices that are needed to create durable feelings of homeliness.

What this book has also demonstrated is that participants in tourism platform economies like *Airbnb* are not passively subjected to a "system" of capital extraction and dispossession. They actively respond to platforms' reconfigurations of everyday life by making them their own, or, by resisting them (see also Leszczynski (2019). Home in the platform economies of tourism is not a site of belonging for either hosts or guests, but rather an ideal or an aspiration for both. The boundaries between shared and intimate spaces are re-negotiated between *Airbnb* hosts and their guests during each new stay. At the conceptual level, this book proposes that home, through its remediation in the *Airbnb* economy, can be understood as a performative site where certain practices constantly challenge feelings of belonging and ontological safety. By conceptualizing home via a performative approach, essentialized ideas of home and travel (e.g., going away from home to have a holiday) have been challenged. By understanding "home" as a set of practices, home is continuously redefined.

In other ways, *Airbnb* has the potential to unsettle and rearrange the social relations between those already living in that "home". In many instances it has induced a new division of labour within the respective households. The extra labour that is needed to accommodate *Airbnb* guests is often regarded by hosts in ways similar to historical gendered, classed, and racialized ideologies pertaining to housework and hospitality labour. In many instances, housework in the *Airbnb* economy, is outsourced to lower-classed and racialized women who oftentimes already carried out this work before the owners of these homes started to rent them out on *Airbnb*, as was the case for Freya (the protagonist in the introduction) and many other hosts interviewed for this research. Yet, these workers do not necessarily reap the profits that these new platform economies generate (for a further development of this argument see Roelofsen and Goyette (2022). Whether outsourced or not, the labour of care upon which the existence of every home must rely, is critically revalorized through both the guests' willingness to pay to stay in that home as well as by their related reviews and ratings. In this way *Airbnb* may indirectly contribute to a substantial transformation and renegotiation of the social relations among those living in that specific home. Home, when incorporated by the *Airbnb* platform economy, thus also represents an opportunity for hosts to produce and extract new values from their intimate spatialities.

Finally, the book has investigated, from a biopolitical perspective, *Airbnb*'s impact at the scale of the everyday life of individuals. By looking through a biopolitical lens and using a qualitative analysis of the platform's digital infrastructure, it has unravelled the ways in which individuals are rated and ranked into a digital Self. Moreover, it has explored the related influence that metrics have on users' everyday lives and how they manoeuvre themselves through and against reconfigurations made by the platform. This is where the book critically reflects on the (bio)political dimensions of datafication by the platform, a reflection that the empirical case has supported in various ways. Through a variety of web applications, *Airbnb* collects and processes information on the everyday lives of individuals in order to determine their position (or "ranking") in a global community of *Airbnb* users. These systems of measurement and the resulting metrics have been designed around a specific set of ideas around "hospitality", "community" and "everyday life" of its affiliates and, accordingly, about what is a "proper" way to manage and opti-mize these very lives on the platform in successful ways. *Airbnb*'s hosting standards are in fact incorporated into the platform's review and rating systems which enable users to qualify each other's performances along a range of "elements". These elements concern the material, aesthetical and locational aspects of a host's home, but also the host's and guest's caring and affective labour that goes into the production of a "homely" and "local" experience to be enjoyed by both users. The hosting and guesting standards in turn provide the platform with parameters for comparison to monitor and the shape the choices and behaviour of *Airbnb* users in very specific ways and in line with the main objectives of the company.

By producing and reconfiguring a "bio-geo-metrics" of its users, the platform has thus the power to rank individuals in the *Airbnb* commu-nity as lower or higher (or even disqualify them from the community) and influence in a crucial way their future transactions. Whether or not indi-viduals "rank well" depends to a considerable extent on their willingness to comply with these (shifting) standards and on their efforts to continu-ously improve their performances according to the changing (bio)politics of the platform. Yet, disqualification from the *Airbnb* community also occurs in other, more direct and abusive ways, enabled by the platform's architecture. Accounts of prejudice towards would-be guests and blatant discrimination were present in the interviews with Freya and numerous other hosts who would outrightly reject booking requests from those who were not white, who had profile photos that supposedly reflected a lower socio-economic status, or whom they expected to have certain religious and political beliefs not matching their own. A critical and biopo-litical reading of the workings of *Airbnb*'s data architecture and interface in this book has thus contributed more evidence on the role of digital platforms in reproducing unfair treatment; excluding and marginalizing users from participating in its very "communities" it promotes to be "diverse and inclusive". It also shows the compliancy of digital plat-forms in broader patterns of discrimination, exclusion and segregation that persist and prevent equal accessibility to recreation, leisure or travel today (see Benjamin and Dillette 2021).

Finally, there are many questions that have been left unanswered in this book and a few recommendations for future research can be made. Although it remains pivotal to examine the impact of tourism-related platforms on large urban contexts receiving high numbers of tourists, the relevant literature should possibly widen its scope to include places that, despite their size and limited popularity, are (or could be) similarly affected by tourism gentrification and socio-spatial segregation. In the case of Bulgaria, where mass tourism in the coastal regions still prevails, cultural tourism is now seen as a key niche of economic development. Bigger numbers of incoming tourists to cultural and historical hotspots in the interior are considered a highly desirable development by the national and local authorities. It remains to be seen how such a push for cultural tourism affects the further development of the "platform economies" of tourism. Case studies like the one presented in this book are necessary to understand the nuances and the influence of the "*Airbnb* effect" on urban (and rural) policy in order to prevent unwanted tourism gentrification.

Chapter 4 has shown that the political and historical specificities as well as the materialities of people's homes crucially shape how people subjectively experience home. It has also shown how hosts relate differently to their homes as soon as they start renting out through the *Airbnb* platform. Whereas the majority of existing studies on *Airbnb* tends to centre on the scale of the city, far less attention has been given to the *Airbnb effect* on the level of the household or the individual. Therefore, more situated accounts of the everyday experiences of *Airbnb* hosts and guests in different contexts could complement the growing body of scholarly work on the broader spatial impact of *Airbnb* on the urban context. Furthermore, an avenue for future research worth further exploring is the agency that platform users have in shaping how their actions are governed by the platform through (everyday) acts of resistance. Chapter 5 has shown how different forms of political activism and opposition have emerged on the platform's fora and blogs that broadly reflect a bottom-up opposition to the ways in which the platform is operationalized. In particular, more research remains to be done on the users' knowledge and awareness of algorithmic decision-making processes and on how such processes are or can be contested. This also relates to the use of data to monitor, control and shape users' behaviour on platforms. While this book has made clear that *Airbnb* hosts and guests (must) create and circulate an amalgam of intimate data about themselves and their homes through the platform's systems, it begs questioning how—in the political economy of data—such practices of accumulation and appropriation can be reformed and regulated, countering its current purpose of selectively benefitting capital interest. Chapter 5 has already provided a critical reading of the ways in which the platform harvests data on its users, but more research could be done on the ethics of such data collection and data repurposing. A critical inquiry into the question of ownership of *Airbnb* data is not only timely but also necessary.

REFERENCES

Airbnb. 2022. "Hosting Makes Airbnb, Airbnb."

Benjamin, Stefanie, and Alana K. Dillette. 2021. "Black Travel Movement: Systemic Racism Informing Tourism." *Annals of Tourism Research* 88 (May): 103169. https://doi.org/10.1016/j.annals.2021.103169.

Blunt, Alison, and Robyn Dowling. 2006. *Home*. Abingdon: Routledge. https://doi.org/10.4324/9780429327360.

Federici, Silvia. 1975. *Wages Against Housework*. Bristol, UK: Falling Wall Press.

Goyette, Kiley. 2021. "'Making Ends Meet' by Renting Homes to Strangers." *City* 25 (3–4): 332–54. https://doi.org/10.1080/13604813.2021.193 5777.

Leszczynski, Agnieszka. 2019. "Glitchy Vignettes of Platform Urbanism." *Environment and Planning D: Society and Space* 38 (2): 189–208. https://doi.org/10.1177/0263775819878721.

Roelofsen, Maartje. 2021. "Capitalizing on Crises. Transformations in Airbnb from the Great Recession Through the Covid-19 Pandemic." In *From Overtourism to Undertourism: Sustainable Scenarios in Post Pandemic Times*, edited by Valentina Pecorelli, 33–54. Milan: UNICOPLI.

Roelofsen, Maartje, and Kiley Goyette. 2022. "Second Shift 2.0. Intensifying Housework in Platform Urbanism." In *Platformisation of Urban Life. Towards a Technocapitalist Transformation of European Cities*, edited by Anke Strüver and Sybille Bauriedl. Transcript Verlag.

INDEX

© The Author(s), under exclusive license to Springer Nature
Switzerland AG 2022
M. Roelofsen, *Hospitality, Home and Life in the Platform Economies
of Tourism*, https://doi.org/10.1007/978-3-031-04010-8